PRAISE FOR *SUPER GENES*

"In *Super Genes*, Drs. Deepak Chopra and Rudolph Tanzi illustrate the interplay of nature and nurture using cutting-edge genetic science and argue persuasively that adapting one's lifestyle can maximize the potential to transcend the inherited susceptibilities handed down to us from our parents."

—James Gusella, Ph.D., director, Center for Human Genetic Research, Massachusetts General Hospital

"Once thought to be the domain of genes, the control of health and behavior is now dynamically linked to the environment and, more important, *our perception of the environment. Super Genes*, by Deepak Chopra and Rudy Tanzi, is a paradigm-shattering synthesis of epigenetic science that offers an easy-to-understand explanation of the mechanisms by which consciousness and environment control our genetic activity. Drs. Chopra and Tanzi's contribution is a valuable resource that empowers us to become the masters of our fate rather than the 'victims' of our heredity."

—Bruce H. Lipton, Ph.D., epigenetic scientist and bestselling author of *The Biology of Belief, Spontaneous Evolution,* and *The Honeymoon Effect*

"The concept that biology is destiny is one of the most pathological and toxic exaggerations to emerge during the entire scientific era. This hard-core materialistic view has been a kind of psychological enslavement that has pushed many people into nihilism and despair. In *Super Genes,* Drs. Deepak Chopra and Rudolph E. Tanzi discuss new evidence that our genes are not our masters, but they respond in large measure to our choices and behaviors. The resulting view honors not just the body but the mind and spirit as well—a vision that is as bright and hopeful as the old view was morbid and depressing. *Super Genes* is an important book. It will empower anyone who reads it, because it expands our view of what it means to be human."

—Larry Dossey, M.D., author of *One Mind: How Our Individual Mind Is Part of a Greater Consciousness and Why It Matters*

"*Super Genes* demolishes the myth that our genes determine our fate. Deepak Chopra and Rudy Tanzi explain in breathtaking detail the magic of how our diet, our lifestyle, our thoughts, and even our gut bacteria or microbiome 'talk' to our genes, regulating which genes get turned on or off, or turned up or down, influencing every aspect of our health. This is essential reading for anyone interested in turning on their health, weight loss, happiness, and longevity genes!"

—Mark Hyman, M.D., director, Cleveland Clinic Center for Functional Medicine, and author of the #1 *New York Times* bestseller *The Blood Sugar Solution*

"We used to think everything about us was either our genetics or our environment. But in *Super Genes*, Deepak Chopra and Rudi Tanzi adeptly teach us that it's all about both—how tightly they are intertwined. And what we can do about it."

—Eric Topol, M.D., author of *The Patient Will See You Now*, and professor of genomics, the Scripps Research Institute

"I have always been far more interested in how we can all optimize our health, as opposed to simply preventing disease. Both are important, no doubt, but teaching people how they can be better—better, faster, stronger, happier—is so much more inspiring. It is what I loved about *Super Brain*, the first book Deepak and Rudolph wrote, and they now have a muscular follow-up with *Super Genes*. In many ways, *Super Genes* is the prequel to *Super Brain*, because it peers down into the very essence of who we are, what comprises us as human beings, and how much of what we experience is preordained destiny vs. being in our own control. The answer to these questions will inspire you.

"We cannot be content to simply blame our genes, but to realize that we can control this blueprint for life and the way our body interprets it.

"Flawlessly weaving together the complicated science of genetics with the touching stories of very real people, my friends Deepak and Rudolph have written a book that you won't put down. You will find yourself scribbling furious notes and sharing your new wisdom with the people you love. First they gave us all the ability to have Super Brains, and now they have done the same with our Super Genes."

—Sanjay Gupta, M.D., neurosurgeon and author of *Chasing Life, Cheating Death,* and *Monday Mornings*

"A groundbreaking and eye-opening account of recent discoveries in two new fields—epigenetics and microbiomics—weaved with practical insights to optimize our own wellness and longevity. Rudy Tanzi and Deepak Chopra, renowned pioneers in their respective fields, have written one of the most important health books of the year."

—Murali Doraiswamy, M.D., professor of psychiatry and medicine, Duke University

"*Super Genes* will take you on an exciting journey of discovery about the ways genetic expression can be modified by simple lifestyle changes and even by how you use your mind. The essential message of this important book is that your genes alone do not determine your destiny. You can learn how to influence them to enjoy better health and optimum well-being. I recommend it."

—Andrew Weil, M.D., author of *Healthy Aging* and *Spontaneous Happiness*

"Our genes are a predisposition, but they are not our fate. The biological mechanisms that affect our health and well-being are often extraordinarily dynamic—for better and for worse. When we eat well, move more, stress less, and love more, our bodies often have a remarkable ability to transform and heal. *Super Genes* is a superb contribution to our growing knowledge that mind, brain, genome, and microbiome can act as a single system. Drs. Chopra and Tanzi continue to make pioneering contributions that are bringing integrative medicine into the mainstream. Highly recommended!"

—Dean Ornish, M.D., founder and president, Preventive Medicine Research Institute, and clinical professor of medicine, University of California, San Francisco

"Chopra and Tanzi have written what will be a life-changing book for many. It will completely change your perspective on how our genes influence us and how we can influence them. Well researched, elegant, and engaging, *Super Genes* furthers our understanding of the potential that lies inside all of us. This is a must-read."

—Steven R. Steinhubl, M.D., director, Digital Medicine, Scripps Translational Science Institute

"This book brings you the sanest, most effective way to participate positively in the very evolution of our whole human species! Deepak and Rudy don't just bring you the wonderful news that you are not a victim of your genes, but dive straight into putting you in charge of your own health through easy, simple, inexpensive changes in your lifestyle that will improve your genome as they bring you, and even your unborn descendants, vibrant good health!"

—Elisabet Sahtouris, evolutionary biologist and futurist and author of *Gaia's Dance: The Story of Earth & Us*

"*Super Genes* is a superb contribution to our growing knowledge that mind, brain, genome, and microbiome are a single system. Congratulations to both Rudy and Deepak."

—Keith L. Black, M.D., professor and chair, Department of Neurosurgery at Cedars-Sinai Medical Center, and author of *Brain Surgeon: A Doctor's Inspiring Encounters with Mortality and Miracles*

"Genetics is a two-way street. Drs. Chopra and Tanzi show how the mind can tell the genes to heal the body."

—Stuart Hameroff, M.D., Banner University Medical Center, the University of Arizona

SUPER GENES

Unlock the Astonishing Power of Your DNA for Optimum Health and Well-Being

DEEPAK CHOPRA, M.D.,
&
RUDOLPH E. TANZI, PH.D.

HARMONY
BOOKS • NEW YORK

Published in the United States by Harmony Books, an imprint of the Crown Publishing Group, a division of Penguin Random House LLC, New York.
crownpublishing.com

Harmony Books is a registered trademark, and the Circle colophon is a trademark of Penguin Random House LLC.

Originally published in hardcover in the United States by Harmony Books, an imprint of the Crown Publishing Group, a division of Penguin Random House LLC, New York, in 2015.

Library of Congress Cataloging-in-Publication Data
Chopra, Deepak author.
Super genes : the hidden key to total well-being / Deepak Chopra, M.D. & Rudolph E. Tanzi, Ph.D.
pages cm
ISBN 978-0-8041-4013-3 (hardback) — ISBN 978-0-8041-4015-7 (paperback) — ISBN 978-0-8041-4014-0 (ebook) 1. Self-care, Health—Popular works. 2. Genes—Popular works. I. Tanzi, Rudolph E., author. II. Title.
RA776.95.C495 2012
613—dc23
2015028562

ISBN 978-0-8041-4015-7
eBook ISBN 978-0-8041-4014-0

Printed in the United States of America

Illustrations by Mapping Specialists, Ltd
Cover design by Pete Garceau
Cover illustration by SergeOstroverhoff/iStockPhoto

10 9 8 7 6 5 4 3 2 1

First Paperback Edition

TO OUR FAMILIES, WITH WHOM WE SHARE THE LOVE
THAT MAKES OUR GENES "SUPER"

CONTENTS

GOOD GENES, BAD GENES, AND SUPER GENES

If you want a better life, what would you change first? Almost no one would say "my genes." And with good reason—we've been taught that genes are fixed and unchangeable: What you were born with is what you will keep for life. If you happen to be an identical twin, both of you will have to settle for identical genes, no matter how good or bad they are. The popular notion of fixed genes is part of our day-to-day language. Why are some people gifted with more beauty and brains than the norm? They have good genes. Why, on the other hand, does a famous Hollywood celebrity undergo a double mastectomy without any sign of disease? It's the threat of bad genes, the inheritance of a strong predisposition to the cancer that runs in her family. The public is frightened, and yet the media doesn't really communicate how rare such a threat actually is.

It's time to explode such rigid notions. Your genes are fluid, dynamic, and responsive to everything you think and do. The news everyone should hear is that gene activity is largely under our control. That's the breakthrough idea emerging from the new genetics and also the basis for this book.

A café jukebox may stand in the corner and never move, but it still plays hundreds of songs. The music of your genes is similar, constantly producing a vast array of chemicals that are encoded messages. We are just discovering how powerful these messages are. By focusing on your own gene activity through conscious choices, you can

Improve your mood level, staving off anxiety and depression
Resist yearly colds and flu
Return to normal sound sleep
Gain more energy and resist chronic stress
Be rid of persistent aches and pains
Relieve your body of a wide range of discomforts
Slow the aging process and potentially reverse it
Normalize your metabolism—the best way to lose weight and keep it off
Decrease your risk of cancer

It was long suspected that genes could be involved when bodily processes go wrong. We now know that genes are definitely involved in making them go right. The entire mind-body system is regulated by gene activity, often in surprising ways. The genes in your intestines, for example, are sending messages about all kinds of things that would apparently have nothing to do with a function as mundane as digestion. These messages concern your moods, the efficiency of your immune system, and your susceptibility to disorders closely related to digestion (e.g., diabetes and irritable bowel syndrome), but also those very distantly related, such as hypertension, Alzheimer's disease, and autoimmune disorders from allergies to chronic inflammation.

Every cell in your body is talking to many other cells via genetic messages, and you need to be part of the conversation. Your lifestyle leads to helpful or harmful genetic activity. In fact, the actions of your genes can potentially be altered by any strong experience throughout your life. So identical twins, despite being born with the same genes, show extremely different gene expression as adults. One twin may be obese, the other lean; one may be schizophrenic and the other not; one may die long before the other. All of these differences are regulated by gene activity.

One reason we called this book *Super Genes* is to raise the bar for what you expect your genes to do for you. The mind-body connection isn't like a footbridge connecting two banks of a river. It's much more like a telephone line—many telephones lines, in fact—teeming with messages. And each message—as tiny as drinking orange juice in the morning, or eating an apple with the peel on, or lowering the noise level at work, or taking a walk before bedtime—is being received by the entire system. Every cell is eavesdropping on what you think, say, and do.

Optimizing your gene activity would be reason enough to throw away the self-defeating notion of good genes versus bad genes. But in reality, our understanding of the human genome—the sum total of all your genes—has vastly expanded over the last two decades. After almost twenty years of research and development the Human Genome Project ended in 2003 with a complete map of the 3 billion chemical base pairs—the alphabet of the code of life—strung along the double helix of DNA in every cell. Suddenly human existence is headed for totally new destinations. It's as if someone handed us a map of an undiscovered continent. In a world where we think there's little left to explore, the human genome is a new frontier.

Let us impress upon you how expanded the field of genetics really is today: You possess a super genome that extends almost infinitely beyond the old textbook ideas of good and bad genes. This super genome is made up of three components:

1. The roughly 23,000 genes you inherited from your parents, together with the 97 percent of the DNA that is located between those genes on the strands of the double helix.

2. The switching mechanism that resides in every strand of DNA, allowing it to be turned on or off, up or down, the way a dimmer switch turns the lights up and down. This mechanism is controlled principally by your *epigenome,* including the buffer of proteins that encloses DNA like a sleeve. The epigenome is as dynamic and alive as you are, responding to experience in complex and fascinating ways.
3. The genes contained in the microbes (tiny microscopic living organisms like bacteria) that inhabit your intestine, mouth, and skin, but primarily your intestine. These "gut microbes" vastly outnumber your own cells. The best estimate is that we harbor 100 trillion gut microbes, comprising between 500 and 2,000 species of bacteria. They are not foreign invaders. We evolved with these microbes over millions of years, and today you wouldn't be able to healthily digest your food, resist disease, or counter a host of chronic disorders from diabetes to cancer without them.

All three components of the super genome are you. They are your building blocks, sending instructions throughout your body at this very minute. You cannot grasp who you are, in fact, without embracing your super genome. How super genes got together to form the mind-body system constitutes the most exciting exploration in present-day genetics. New findings are emerging in a flood of knowledge that affects all of us. It's changing the way we live, love, and understand our place in the universe.

The new genetics can be simplified in a single phrase: *we are learning how to make our genes help us.* Instead of allowing your bad genes to hurt you and your good genes to give you a break in life, which used to be the prevailing view, you should think of the super genome as a willing servant who can help you direct the life you want to live. You were born to use your genes, not the other way around. We aren't indulging in wish fulfillment here—far from it. The new genetics is all about how to alter gene activity in a positive direction.

Super Genes gathers the most important findings we have today and then expands upon them. We combine decades of experience as one of the world's leading geneticists and one of the world's most acclaimed leaders in mind-body medicine and spirituality. We may come from different worlds, and we spend our working days in divergent ways, Rudy doing cutting-edge research into the cause and potential cure of Alzheimer's disease, Deepak teaching about mind, body, and spirit to hundreds of audiences a year.

However, we're united in a passion for transformation, whether the roots of change are found in the brain or in the gene. Our previous book, *Super Brain*, used the best neuroscience to show how the brain can be healed and renewed, optimizing its daily function to create much better outcomes in people's lives.

Our new book deepens the story—you could call it a prequel to *Super Brain*—because the brain depends on the DNA in every nerve cell to do the amazing things it does every day. We are taking the same message—you are the user of your brain, not the other way around—and extending it to the genome. Lifestyle is the domain where transformation takes place, whether we're talking about super brain or super genes. There is the possibility, through simple lifestyle changes, of ending up as a person who is activating an enormous amount of untapped potential.

The most exciting news of all is that the conversation between body, mind, and genes can be transformed. This transformation goes far beyond prevention, even beyond wellness, to a state we call radical well-being. This book explains every aspect of radical well-being, showing how up-to-date science either totally supports it or strongly suggests what we should be doing if we want the most life-supporting response from our genes.

The terms *good genes* and *bad genes* are misleading because they feed into a bigger misconception: biology as destiny. As we'll explain, there are no good versus bad genes. All genes are good. It is *mutation*—variations in the DNA sequence or structure—that can turn genes bad. Other mutations can also turn genes "good." Disease-associated gene mutations that will actually destine a person to acquire a disease with certainty in the span of a normal life span amount to only 5 percent of all disease-associated mutations. This is a minuscule portion of the three million or so DNA variations in each person's super genome. As long as you keep thinking in terms of good genes and bad genes, you've imprisoned yourself in bad, outmoded beliefs. Biology is being allowed to define who you are. In modern society, where people have more freedom of choice than ever before, it's ironic that genetics became so deterministic. "My genes did it" became the blanket answer to why someone overeats, suffers from depression, breaks the law, has a psychotic break, or even believes in God.

If the new genetics is teaching us anything, it's about nature cooperating with nurture. Your genes can predispose you to obesity or depression or type 2 diabetes, but this is like saying that a piano predisposes you to play wrong notes. The possibility exists, yet far more important is all the good music a piano—and a gene—are capable of.

We offer you this book in the spirit of expanding your wellbeing, not because there are so many wrong notes to avoid, but because there's so much beautiful music left to be composed. Super genes hold the key to personal transformation, which has suddenly become far more attainable—and desirable—than ever before.

WHY SUPER GENES?

An Urgent Answer

The purpose of this book is to raise everyday well-being to the level of radical well-being. Such a goal requires a journey of transformation through an understanding of our own genetics. This fascinating field of inquiry has led to a flood of exciting findings, and more appear every day. Human DNA has many more secrets to reveal. Yet a tipping point has already been reached. It has become blindingly clear that the human body is not what it seems to be.

Imagine you are standing in front of a mirror: what do you see? The obvious answer is a living object, a moving machine of flesh and blood. This object is your home base and protective shelter. It faithfully takes you where you want to go and does what you want to do. Without a physical body, life would have no foundation. But what if everything you assumed about your body were an illusion? What if that *thing* you see in the mirror isn't a thing at all?

In reality, your body is like a river, constantly flowing and changing.

Your body is like a cloud, a swirl of energy that is 99 percent empty space.

Your body is like a brilliant idea in the cosmic mind, an idea that took billions of years of evolution to construct.

These comparisons aren't just images—they are realities pointing to transformation. Right now, the body as a physical thing fits in with everyday experience. To paraphrase Shakespeare, if you cut yourself, do you not bleed? Yes, of course, because the physical side of life is totally necessary. But the physical side comes second. Without those other possibilities—the body as idea, energy cloud, and constant change—your body would fly away, vanishing into a random swirl of atoms.

Once you see past the facade of that image in the mirror, the big story begins. Behind the mirror, so to speak, genetics has been unfolding the story of life in stages, punctuated by the 1953 breakthrough that revealed DNA's double helix, a twisted ladder with billions of chemical rungs. In the past ten years, however, the story has exploded, thanks to the discovery of how active our genes really are. Everywhere in the body, a cell puts the secret of life into practice:

It *knows* what's good for it and seizes upon the good.
It *knows* what's bad for it and avoids the bad.
It sustains its survival from moment to moment with total focus.
It monitors the well-being of every other cell.
It adapts to reality without resistance or judgment.
It draws upon the deepest resources of Nature's intelligence.

Can we, the summation of all those cells, say the same for ourselves? Do we eat too much, overindulge in alcohol, put up with pummeling stress, and rob ourselves of sleep? No healthy cell would make such choices.

So why the disconnect? Nature designed us to be as healthy as our cells. There is no reason not to be. Cells naturally make the right choices at every moment. How can we do the same?

What's so exciting about recent research is that gene activity can be greatly improved, and when this happens a state of radical well-being is possible. What makes it radical is that it goes far beyond conventional prevention. The very foundation of chronic disease is being exposed by the new genetics. We are seeing how lifestyle choices made years ago profoundly affect how the body operates today, for both good and ill. Your genes are eavesdropping on every choice you make.

We hold that radical well-being is an urgent need, and we believe wholeheartedly that we can convince you of this. Unknown to the vast majority of people, there's a hole in conventional well-being, a hole big enough that accelerated aging, chronic disease, obesity, depression, and addiction have managed to slip through. All efforts to counter these threats have been only half successful at best. A new model is needed. Here's how one woman experienced this need.

RUTH ANN'S STORY

When Ruth Ann developed pain in both hips, she initially shrugged it off. At fifty-nine, she prided herself on how well she was managing her body. She had superb impulse control, eating the right foods without the snacking and guilty dashes to the fridge for ice cream at midnight that gradually put on pounds. She didn't smoke and rarely drank. Her cupboard held a stock of vitamins and nutritional supplements. Her exercise routine went beyond the recommended minimum of four or five periods of vigorous activity per week—she spent two hours at the gym every day. As a result, on the eve of turning sixty, Ruth Ann could show off a perfect figure, which had been her main focus all along.

The arrival of pain in her hips two years earlier was annoying, but she didn't let it affect her exercise routine. Gradually the pain became chronic; it spiked whenever she ran on the treadmill. Eventually she needed to lie down for an hour every afternoon to allow the pain to subside. Ruth Ann went to her doctor. X-rays were taken, and the news was bad: She had degenerative osteoarthritis. Sooner or later, the doctor informed her, she was facing a hip replacement.

The cause of arthritis, of which there are many types, is unknown, but Ruth Ann has her own explanation. "I shouldn't have been such an exercise fanatic. I pushed myself too hard, and now I'm paying the price." She felt defeated. In her mind, she had been doing all the right things to postpone "turning into an old lady." This was her biggest fear. Now, as if tiny goblins were coming out of the closet, the symptoms of accelerated aging were upon her. Her figure is that of a thirty-year-old, but appearances deceive. She feels tired for no reason. Her sleep and appetite have turned irregular, with nights of severe insomnia that can go on for several weeks. Small stresses give rise to low-level anxiety. Ruth Ann has never felt helpless before. Whenever she has a mental image of herself as an "old lady," she wishes she could run back to the gym and get on the treadmill again.

The bottom line is that Ruth Ann feels her body has betrayed her. Yet consider how the situation looks from a cell's point of view. A cell doesn't push itself beyond its limits. It heeds the slightest sign of damage and rushes to repair it. A cell obeys the natural cycle of rest and activity. It follows the deep understanding of life embedded in its DNA. By conventional standards, Ruth Ann did all the right things, yet at a deeper level she was disconnected from her body's intelligence.

We have so much that's positive to tell you that we will state the negative side just once: The two major threats to well-being—illness and aging—are constantly present. Out of sight, without your knowing it, your present good health is being silently undermined. Abnormal processes are taking place in everyone's body at a microscopic level. Anomalies inside a cell that affect only a cluster of molecules or the shape of one enzyme are virtually undetectable. You can't feel them as an ache or pain or even as vague discomfort. Such abnormalities can take years to develop into even minor symptoms.

But the day will arrive when our body starts to tell us a story we don't want to hear, just as Ruth Ann's body did.

This book tells you how to avert that day for years, or even decades, to come. The possibility of radical well-being is very real, and the most exciting developments are merely a prelude to a revolution in self-care. Become a pioneer in that revolution. It's the most significant step you can take in shaping the future you desire for body, mind, and spirit. Your genes play a part in all of these areas, as we're about to show you.

FROM GENES TO SUPER GENE

The threats that undermine your well-being are persistent. Even if you consider yourself safe right now, how secure is your future? Genes can help answer that question. They can lead you to make life-supporting choices while correcting the wrong choices made in the past. The first step is to focus on the cell. Your body has approximately 50 trillion to 100 trillion cells (estimates vary widely). There is no process—from thinking a thought to having a baby, from fending off invading bacteria to digesting a ham sandwich—that isn't tied to a specialized activity in your cells. A cell must look to its DNA to keep it perfectly functioning, because DNA, as the "brain" of the cell, is ultimately in charge of every process. In a healthy person, this activity occurs perfectly more than 99.9 percent of the time. It's the tiny exceptions, amounting to the merest fraction of 0.1 percent, that can cause trouble.

The DNA that's neatly tucked inside each cell is something magnificent, a complex combination of chemicals and proteins that holds the entire past, present, and future of all life on our planet. Bacteria are essential to the body, too, with trillions of them lining the gut and the surface of the skin. These form colonies known as the microbiome. It's long been known that bacteria in the intestines make digestion possible. But recently the microbiome has assumed much greater importance. For one thing, there's the sheer number of bacteria involved, which amount to something like 90 percent of the cells in the body. Even more crucial, bacterial DNA became part of human DNA over the course of billions of years. It is estimated that 90 percent of the genetic information inside us is bacterial—our ancestors were microbes, and they are, in many ways, still present in the structure of our cells.

In fact, your body may contain 100 trillion or more bacteria (a very rough estimate). In isolation, they would weigh somewhere between three and five pounds in dry weight. If we keep score by the number of different genes you possess, it would be about 23,000 genes inside your cells and 1 million genes for all these various microbes. In a sense we are sophisticated hosts for the micro-organisms that colonize us. The implications for medicine and health are potentially staggering and are just now being explored. One conclusion is inescapable: the human genome, having expanded tenfold, has become a super genome. Because of the microbes now being wrapped into the story, Earth's 2.8-billion-year-old genetic legacy is present inside each of us, here and now. Much of the original stuff, genetically speaking, is still propagating inside the cells of your body.

The fact that DNA stores the entire history of life gives it tremendous responsibility. One slip, and an entire species can be wiped out. Realizing this fact, geneticists spent many decades thinking about DNA as a stable chemical, its biggest threat being the instability created when a mistake slips by the body's defenses. But now we realize that DNA is responsive to everything that happens in our lives. This opens the door to many new possibilities that science is just now beginning to grasp.

SASKIA'S STORY

Some people find themselves apparently victimized by their genes; others are rescued by them. One woman experienced both. Saskia is in her late forties with advanced breast cancer that has metastasized to other locations in her body, including her bones. In her most recent battle against the disease, Saskia bypassed chemotherapy in favor of immunotherapy, which aims at increasing the body's own immune response. She also decided to spend a week learning how to take care of herself through meditation, yoga, massage, and other complementary therapies. (The program she attended was given at the Chopra Center. We mention this in the spirit of full disclosure, not to take credit for what occurred next.)

Saskia enjoyed the week and came away with a feeling that she could relate to her body in a better way. She appreciated how well she was treated, pointing in particular to the loving attitude of the massage therapists. At the end of the week she reported that her bone pain had gone away, and she went home feeling much better, emotionally and physically. She recently sent a follow-up e-mail describing what happened next.

> The day after I got home, I had another PET/CT scan. This one was four months after the last. The following week I met with my oncologist. Though I was expecting the worst, I had decided that no matter how bad my scan looked, I felt a lot better, and that's what counted. But instead of bad news, he told me that he had never seen such a response in such a short time, and especially without the use of chemo drugs. . . . He was very surprised and is much more interested now in what I'm doing!
>
> I told him about what I learned at the Chopra Center (especially meditation, yoga, and massages), the dietary changes I'd made, and how supportive my husband has been in these last few months. I believe that all these things were working together to make healing possible.
>
> Basically all the many metastases to my lymph nodes are gone, as well as the metastases to my liver; more than half of

> the mets to my bones have disappeared. The remaining bone mets have all diminished greatly in size. There's one new lymph node met on the left side of my neck, but the doctor believes it's insignificant in light of the vast improvements everywhere else. He told me to just keep doing whatever I'm doing.

There are two attitudes to take to this story. One is the standard medical response, which amounts to dismissal.

Faced with Saskia's experience, most oncologists would consider it merely another piece of anecdotal evidence that has little bearing on the overall statistics relating to cancer treatment and survival. Cancer is a numbers game. What happens to thousands of patients tells the tale, not what happens to one patient. The other attitude to Saskia's experience is to explore how changes in her situation led to such a remarkable result. Let's list all the changes she experienced that might influence gene expression:

Improved attitude toward her disorder
Increased optimism
Decreased bone pain
Emotional support from her husband
New knowledge about the mind-body connection
New lifestyle choices added to her daily routine: meditation, yoga, massage
Benefits from therapeutic massage and other treatments at the center

The list looks quite diverse, and only one or two items on it would be found under current standard cancer treatments. But there's a common thread to every item. New messages were sent to and from her brain and her genes. If medicine could decode these messages, we'd get much closer to solving the mystery of healing. It can be hard for any physician who is in the business of curing his patients to admit that the only true healer is the body itself. And how the body pushes atoms and molecules around to achieve healing—or not—remains a deep mystery.

What will happen to Saskia in the coming months and years is unpredictable. We are not promoting miracle cures in any way, shape, or form. We know full well that *miracle* isn't a useful term for understanding how the body operates.

If you could listen in on the stream of messages received at the genetic level over the course of a single day, in all likelihood you'd hear the following:

Keep doing what you're doing.
Reject or ignore change.
Keep problems away from me. I don't want to know about them.
Make my life pleasant.
Avoid difficulties and pain.
You take care of it. I don't want to.

You aren't aware that this is what you are telling your genes, over and over, because you don't put these messages into words like a telegram. But your *intention* is clear, and cells respond to what you want and do, not what you say. Each of us is incredibly fortunate that our bodies can run automatically with almost total perfection for decades at a time. But unless we participate in our own well-being, sending conscious messages to our genes, running on automatic isn't good enough. Radical well-being requires conscious choices. When you make the right choices, your genes will cooperate with whatever you want.

This is the new story we want you to follow, and to turn into your own story. When you use your genes for transformation, they become super genes. To guide you to the goal, the rest of the book is organized into three parts:

The Science of Transformation: Here we give you the latest knowledge about the new genetics and the revolution that is changing biology, evolution, inheritance, and the human body itself.

Lifestyle Choices for Radical Well-Being: Here we provide a path for change that's both practical and, as much as possible, effortless.

Guiding Your Own Evolution: Here we go to the source of all growth and change, which is consciousness. You cannot change what you aren't aware of, and when you are totally aware, the promise of self-directed transformation comes true.

There's the map. Now we begin the journey. The map has marked out the territory to be covered, but until you enter the territory, it won't become real for you. What makes this journey unique is that every step has the power to change your personal reality. Nothing could be more fascinating or more rewarding.

Almost a thousand years before DNA revealed its first secret, the mystic Persian poet Rumi took the same journey. He looked over his shoulder to tell us where the road leads:

Motes of dust dancing in the light
That's our dance, too.
We don't listen inside to hear the music—
No matter.
The dance of life goes on,
And in the joy of the sun
Is hiding a God.

Part One

THE SCIENCE OF TRANSFORMATION

Thanks to the genetic revolution happening all around us, a new and powerful ally has appeared to aid human happiness. The notion that DNA contains the code of life isn't new. But it's very new to say that you can *use* your genes. DNA isn't locked up like a frozen bank account you can't draw on. As we mentioned earlier, the old belief that "biology is destiny" no longer has the iron grip it once did. The science of transformation tells a new story, of endless possibilities arising from DNA. But to understand that story, we need to look at DNA in all its fantastic complexity.

The evolution of all planetary life is condensed inside deoxyribose nucleic acid, to use DNA's full name. A single strand of DNA is 3 meters long, yet it fits into a space of only 2 to 3 cubic microns in the cell's nucleus (1 micron = 1 millionth of a meter, or roughly 1 millionth of a yard). Only about 3 percent of your DNA is made up of genes, which provide the blueprints for proteins and ribonucleic acid (RNA), the facsimile of DNA with which proteins are made or gene activity can be regulated. These, together with fat, water, and a huge host of friendly microbes, make up your physical body. To a geneticist, you are a highly complex colony built by DNA, and you are constantly being rebuilt.

The body's superstructure is constantly under revision based on how you live your life. What's known as gene expression—the thousands of chemical products produced by genes—is highly malleable. This goes against what most people know or believe. For instance, how many times have you heard these common phrases: "he's a chip off the old block"; "the apple doesn't fall far from the tree"; "he's just like his old man"? Just how true are the old adages? Are we really just the repeat biology and continuing personality of our parents, with a few variations thrown in?

The new genetics says no. Like your brain, which responds to every choice you make, your genome is constantly responsive. While the genes your parents passed on to you won't change into new genes—your unique blueprint stays the same throughout your lifetime—gene activity changes fluidly and often very quickly. Genes are susceptible to adverse change that can occur as the result of diet, disease, stress, and other factors. That's why everyday lifestyle choices have repercussions down to the genetic level. It's entirely through gene expression that the body's intelligence acquires physical form. What's even more astonishing, as we will see, is that how you influence your body today may be felt in the well-being of your children and grandchildren far in the future.

Besides DNA, your genome is made up of special proteins that support and "cushion" the DNA. DNA itself is composed of four chemical bases that pair up to form rungs on the double helix.

These four bases are adenine (abbreviated as A), thymine (T), cytosine (C), and guanine (G). The fact that an alphabet of only four letters is responsible for every life form on Earth never ceases to astonish. Here's how complexity arises from simplicity: A pairs with T, and C pairs with G. Your unique genome carries 3 billion of these bases from each parent. The 3 billion bases are divvied up into 23 chromosomes, which are labeled from 1 to 22 plus the sex chromosomes, X and Y. The mother always gives her baby an X chromosome. If the father gives a Y chromosome, the baby's sex will be male; if an

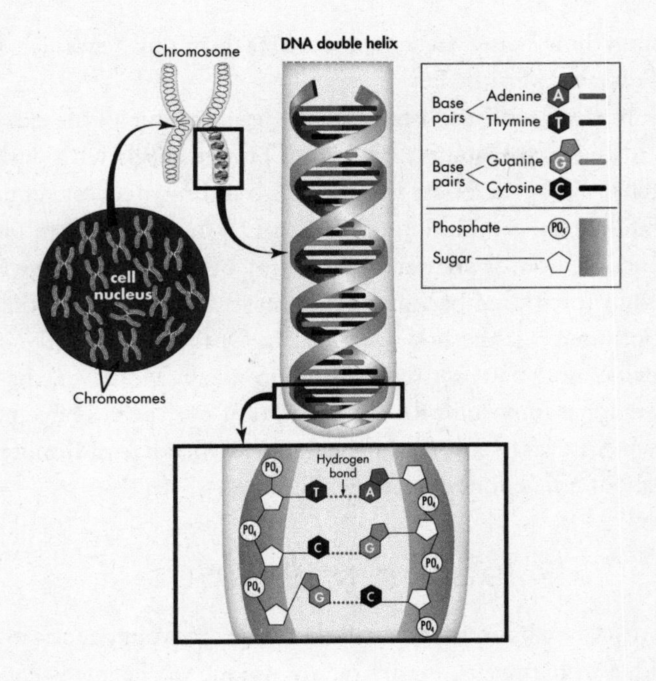

X chromosome, the sex will be female. Since each of your parents gave you 23 chromosomes and 3 billion bases of DNA, your cells contain a total of 46 chromosomes and 6 billion bases. It's possible to see already how Nature supplied itself with enough building materials to make a moth, a mouse, or a Mozart out of four letters.

The completion of the epoch-making Human Genome Project in 2003, along with subsequent studies, yielded some surprising, even baffling, results. For example, our genome contains roughly 23,000 genes, which is far fewer than anyone supposed. We consider *Homo sapiens* the most evolved life-form on Earth, but that's not the same as having more genes—the genome of rice, which contains only 12 pairs of chromosomes, has as many as 55,000 genes! How as a species do we get away with fewer genes than a grain of rice? The answer has to do with how efficient our genes have become, and especially how many diverse proteins each of our genes can make. Gene expression is the key.

Compared with the genes in rice, each of our genes can make many different versions of the same protein, each with a slightly different role in the body, whether it's building a cell or regulating it. Thanks to the evolution of human DNA, we get more biological function from fewer genes. Economy of scale, together with redundancy (providing backup so that survival doesn't depend on one genetic system), is the rule in evolution. Our genes are still evolving to provide more bang for the buck, so to speak. Moreover, the genes that are most important for the survival of our species have backup copies just in case some become corrupted with harmful mutations. Talk about efficient and forward thinking!

BECOMING UNIQUE

From just these basic facts, it becomes clear that your genetic makeup is unique in two ways. First, you are unique in the genes you were born with, which no one else duplicates unless you are an identical twin. Second, you are unique in what your genes are doing right at this moment, because this activity is your story, the book of life that you are the author of. The outcome of ordinary lifestyle choices (*Do I go to the gym or stay home? Do I gossip at work or stay out of other people's business? Do I donate money to charity or fatten my bank account instead?*) depends on a single question: *What am I asking my genes to do?* The back-and-forth between you and your genome is the determining factor in your present and future.

It doesn't take the whole genome to make you unique, however. In the three billion bases of DNA that each parent give you, there is a difference once every thousand bases compared with the vast majority of human DNA on the planet. This means that each of your parents passed on roughly three million bases that are known as DNA variants. A DNA variant can sometimes, but rarely, guarantee a certain disease within a normal life span or simply serve to increase one's risk without guaranteeing the disease. For example, at one of the 3 billion steps of the double helix, you may have the base A while your sibling has a T. This difference may result in your being predisposed to developing a disease like Alzheimer's or a particular form of cancer, whereas your sibling is not.

Contrary to public perception, there is no such thing as a "disease gene." All genes are "good" and provide a normal function needed by the body. It's the variants they harbor that can bring problems. On the positive side, some mutations increase resistance to disease. A few rare family strains, for example, have given almost total immunity to heart disease. No matter how much fatty food is in their diet, the cholesterol isn't converted into blood fats that line the coronary arteries with plaque. Geneticists have sought out these isolated populations to discover which variant might have gifted them with resistance to heart disease. By the same token, there are small, rare populations in which presenile Alzheimer's disease affects almost the entire family line. They, too, must be studied in an attempt to discover if a genetic signature is responsible for such a bad outcome.

Rudy was fortunate to be intimately involved with the earliest pioneering events of the current genetics revolution. When he and his colleague Dr. James Gusella were still in their early twenties, carrying out the first mapping of the human genome at Massachusetts General Hospital, they became the first researchers in the world to locate a disease-causing gene by tracking natural DNA variants in the genome. In their landmark study, they were able to show that the gene for Huntington's disease resides on chromosome 4. Huntington's disease is a fatal disorder in which no clues about the cause were previously available.

Some variants are common ones, being present in more than 10 percent of the human population, while others are rare, isolated mutations. A genetic variant can predispose you to certain diseases or behaviors, which is why research focuses so intensely on the genetic contribution to Alzheimer's or depression. Other variants do nothing at all, at least not so far in our evolution. Your personal DNA "fingerprint" is based on the set of variants you inherited. These determine both the functioning and structure of the hundreds of thousands of different types of proteins in your body.

The number of gene variants that give you a fixed characteristic like blue eyes or blond hair are known as fully penetrant gene variants, and they are in the vast minority, as few as 5 percent of the total. But, in the vast majority of cases regarding health and personality, your genetic destiny is not set in stone. Genes are only one component of the almost infinite interplay of DNA, behavior, and the environment.

This fact was underlined by a 2015 study on autism published in the journal *Nature Medicine*. Autism is a baffling disorder because there is no single kind of autism, but rather a wide spectrum of behavior, one that Rudy has worked on extensively over the course of his career. The mass media image of an autistic child portrays a totally withdrawn state in which the child hardly reacts to any outside stimuli. Totally lost in himself, he may rock back and forth or "twiddle" with repeated, robotic gestures. Emotions are stunted or nonexistent. The parents are desperate to find a way to break through the shell.

But in some families there are two autistic children, and more often than not, the parents say that their behavior is very different. The new study, which looked at the genes of autistic siblings, confirmed this impression. Researchers looked at eighty-five families in which two children had been diagnosed with autism. It's possible, through techniques known as genome-wide association screens and whole genome sequencing, to look at millions of DNA variants in someone's genome. The study targeted 100 specific variants that have been genetically associated with a greater risk of being autistic. To everyone's surprise, only about 30 percent of the autistic siblings shared the same mutation in their DNA, while 70 percent did not.

In the shared group, the two autistic children behaved more or less alike. But in the unshared group, the 70 percent, their behavior was as different as any two brothers or sisters. What this suggests is that autism is unique because each person is unique. Even if scientists examined the genome of thousands and thousands of autistic children, it would be extremely challenging to determine the biological basis of the disease.

Unfortunately, not being able to predict autism in advance brings us back to a state of uncertainty. The chances of having two autistic children in a family of four or more is remote, about 1 in 10,000. As reported in the *New York Times*, a Canadian couple who already had one severely autistic child and one child with no developmental problems went to the doctor's with their wish to have a third child. What was the risk that the new baby would be autistic? Hospitals examine the genome of the oldest affected child to arrive at a prediction. In this case, the couple were told that the chances of having another autistic child were slim, and in any event, if the child were autistic, it wouldn't necessarily be to a severe degree.

But, in reality, the new baby, which the couple decided to have, did develop severe autism. And the couple report that their two autistic children don't behave alike. One is outgoing enough to run up to strangers, while the other holds back. One loves to play with computers; the other has no interest. One runs around, while the other prefers to sit in one place.

This is the outcome of diversity. No matter how many genetic samples you take from a family line, the next baby to be born will be largely unpredictable, not just in terms of the risk of autism but in general.

While genes clearly determine some things, like the onset of some rare forms of disease, most of the time the gene variants that we inherit merely confer a *susceptibility* toward a disease. The same can be said about genetic predisposition to certain behavior or personality types. The bottom line is that what we do, what we experience, and how we view the world, along with what we are exposed to in our environment, strongly influence the actual outcome of the genes we inherit. No one can put a precise number on how much influence you can exert on your gene expression. But there's no longer any doubt that your influence is important, because it's in play all the time.

It's now possible to reconstruct the genome of Neanderthals from their remains, but no matter how minutely their genes are examined, the future evolution of humans isn't observable. There is no gene for mathematics or science. If you compared Mozart's genes to an amateur violinist's, you couldn't detect which one was the musical genius. Even the most basic predictions are turning out to be far from simple. A pregnant mother might want to know how tall her baby will grow up to be. There isn't a single gene for height. So far, it seems that more than twenty genes are involved. Even if you could predict how these twenty genes will express themselves, at best you would arrive at 50 percent of the answer. Environmental factors like diet, including both the mother's diet and the baby's, will contribute the other half.

Let's be extremely generous and foresee that genetics, using some sort of super computer, might one day handle all the interlocking physical factors. With all those data, predicting how tall a child will grow up to be would still remain uncertain, because unexpected events always arise. There is a condition known as psychological dwarfism, for example, in which young children raised in an abusive family situation become stunted in their growth. The mind-body connection has turned a psychological factor, heavily weighted with emotional damage, into physical expression. In short, DNA's alphabet has immeasurable "words" to write, and what they will be is unknown.

Sometimes you can witness in action how life experiences alter a person's DNA. At the end of each chromosome is a section of DNA called a *telomere*, which protects the chromosome from unraveling, like the tip of a shoelace. As we age, our telomeres get shorter with every new division of a cell. After dozens of divisions, the protective telomeres become so short that the cell becomes senescent—that is, it stops being able to divide anymore. The death of the cell follows, along with the absence of new cells to replace it.

As it turns out, a person's experiences also affect telomeres. Scientists at Duke University analyzed DNA samples first from five-year-olds and then again when the children were ten. The researchers knew that some of these children experienced physical abuse, bullying, or violent domestic disputes. The ones who experienced the most negative and stressful experiences underwent the most rapid erosion of their telomeres. On the other hand, other research indicates that exercise and meditation have been shown to increase the length of telomeres.

The implications are profound. Longevity is not only influenced by the DNA variants inherited in select genes from your parents. What happens to you today will perhaps show up tomorrow in the structure of your chromosomes.

One of the most fascinating journeys in the new genetics revolves around life experiences and our genes. Human existence is infinitely complex, which makes it a bewildering task to understand how genes react to daily life. Somehow they do, and we've made a start in revealing how they do it—that's the subject of our next chapter, which exposes many new possibilities and many mysteries at the same time.

HOW TO CHANGE YOUR FUTURE

The Arrival of Epigenetics

What enables genes to be just the opposite of fixed—fluid, malleable, and interconnected—falls under a new field called *epigenetics*. The Greek word *epi* means "upon," so *epigenetics* is the study of what is on top of genetics. Physically, *epi* refers to the sheath of proteins and chemicals that cushion and modify each strand of DNA. The entire amount of epigenetic modification of the DNA in your body is known as the *epigenome*. Research on the epigenome is probably the most exciting part of genetics right now, because it is here that genes get switched on and off (like a light switch) and up and down (like a thermostat). What if we can control these switches voluntarily? The prospect makes any adventurous geneticist dizzy with the possibilities.

In the 1950s, before it was suspected that the epigenome existed, an English biologist named Conrad Waddington first proposed that human development from embryo to senior citizen was not completely hardwired in DNA. It took decades for the notion of genetic "soft wiring" to catch on, for the now familiar reason that genes were thought to be fixed. But eventually it was impossible to ignore certain anomalies. Identical twins are the classical example, because they are born with identical genes. If DNA hardwires them, identical twins should be biologically predestined to be exactly the same all their lives.

But they aren't. Identical twins with virtually the same genomic DNA can be very different based on how they experience the world and how this translates into gene activity. If you know a set of twins, you've no doubt heard them express how different they feel from each other. It takes more than the same genome to create a person. Two identical buildings can be constructed with the same blueprints but be very different places based on the activities inside. Schizophrenia, for example, is known to have a genetic component, yet if one twin is schizophrenic, there is only a 50 percent chance that the other will be. This mystery requires further discussion, but you can see the dilemma posed for "biology as destiny." Epigenetics was born when geneticists focused on the controls behind gene expression. It turns out that the flexibility of these controls is one of the most precious of life's gifts.

While all the cells in your body have largely identical DNA sequences and genetic blueprints, each of the two hundred or so different cell types possesses different structures and roles. Under a microscope, a neuron looks so different from a heart cell that you would hardly expect them to be operated by the same DNA. Genes are programmed to create a variety of different cells from stem cells, which are the "baby" precursors to mature cells. Stem cells stored in your bone marrow, for example, replace your blood cells as they die, which is every few months. The brain has a lifetime supply of stem cells also, which allows for the generation of new neurons at any stage of life—very good news for an aging population that wants to remain as vital and mentally alert as possible.

A complete understanding of "soft" inheritance is now unfolding, and every step brings new surprises. In a 2005 study, Dr. Michael Skinner showed that exposing a pregnant rat to chemicals that impair sexual function resulted in fertility problems in offspring onward down to her great-great-grandchildren. Surprisingly, the fertility issues were transmitted to the next generation as a "soft" inheritance by the male rats—via chemical tags (known as methyl groups) on the DNA—along with the DNA sequence of the parents. We know the transmission was not a "hard" inheritance because the actual DNA sequence of the transmitted genes remained the same.

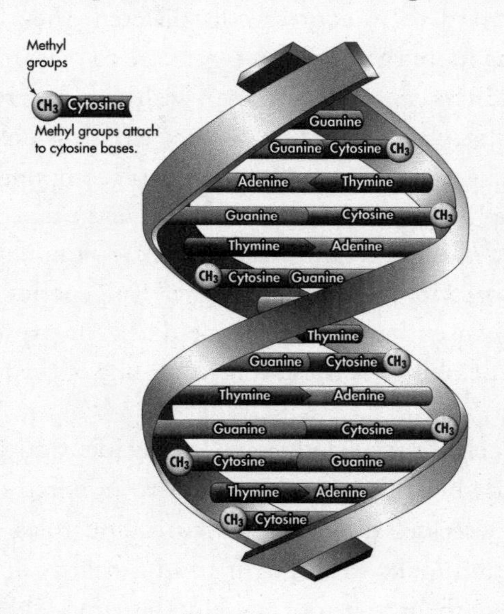

If DNA is the storehouse for billions of years of evolution, the epigenome is the storehouse of short-term genetic activities, both very recent and extending back one, two, or several generations. The fact that memory can be inherited isn't new in biology. The bones in the fins of ancestral fish are the same in structure as the bones in the paws of mammals and those in our own hands. This kind of memory is definitely hardwired, because evolution from species of fish, bears, raccoons, and *Homo sapiens* took millions of years to become fixed. What's new with epigenetics is that the memory of *personal* experience—yours, your father's, your great-grandmother's—may be immediately passed on.

This brings us to probably the single most important idea in the new genetic revolution. The epigenome allows for genes to react to experience. They are not isolated but are open to the world just as much as you are. This offers the possibility that how you react to your daily life, physically and psychologically, can be passed on through soft inheritance. Simply put, when you subject your genes to a healthy lifestyle, you are creating super genes. Such a possibility would have seemed like science fiction in previous eras, when it was set in stone that only DNA is passed down from parents to offspring. But in a landmark study from 2003, scientists took two groups of mice developed with a mutant gene that made them be born with both yellow fur and a voracious appetite. They were thus genetically programmed to overeat to the point of obesity.

The researchers then fed one group of the mice a standard mouse diet, while the other group was given the same food with added nutritional supplements (folic acid, vitamin B_{12}, choline, and the sugar beet product betaine). As it turned out, the offspring of the mice given the supplements grew up with brown fur and normal weight despite the mutant gene. Astonishingly, the mutant gene for yellow fur and voracious appetite was overridden by the diet of the mother. In support of this finding, another study found that mice whose mothers received fewer vitamins were more predisposed to obesity and other diseases. Thus a mother's nutritional state may have a more profound impact on her baby than was previously believed.

The implications of these studies were revolutionary on several fronts. First, the epigenome is always interacting with daily life. What happens to you today is being recorded at the epigenetic level and—if humans react in the same way as mice—is potentially passed on to future generations. Your own predispositions may not belong solely to you, then. They exist on a kind of genetic conveyor belt on which each generation adds its own contribution.

Another study, published in 2005, showed that pregnant women who witnessed the 9/11 attacks on the World Trade Center passed on to their babies higher levels of the stress hormone cortisol. Your mother's or grandmother's traumatic childhood may have changed your own personality toward anxiety and depression. If the genome is the architect's blueprint of life, the epigenome is the engineer, construction crew, and facilities manager all in one.

A DUTCH MYSTERY

We've established how epigenetics delves into the changes in gene activity driven by life experiences. Such changes require no alterations in the DNA sequence itself—that is, no mutations. Some kind of switching mechanism is involved instead, but it's not a simple on or off. The switching mechanism for DNA turns out to be as complicated as human behavior. Think of a common behavior like losing your temper. Anger can flick on and off like switching the lights on and off, or it can simmer for a while. Anger can be hidden from view, disguised by being in control of one's emotions. Once it flares, anger can range from mild to explosive. Everyone accepts these distinctions, since by common experience we all know hotheads and cool customers. In ourselves we know how to swallow our anger, yet at the same time we fight against it.

Now translate this situation into genetic activity, and all the same variables apply. Any activity of a gene can be hidden, or turned off. It can be partially or totally expressed, going up or down as if controlled by a thermostat. And just as anger is intertwined with every other emotion, so is every gene intertwined with every other gene. It's looking truer and truer that any subjective experience owes its complexity to a parallel complexity at the microscopic level.

Where does it leave us to know how much we don't know? If emotions handle genes and genes handle emotions, the circularity might be endless. Having brought us to the control room where all the switching is done, epigenetics still hasn't put the switches in our hands. Mastering the controls is everyone's individual responsibility. Otherwise, genetic changes can be quite drastic when no one is in control. Let's explore a widely publicized and very puzzling example.

Below is a chart of male height in Europe from 1820 to 2013, as compiled by computer science researcher Randy Olson. (There are other calculations that differ from the one pictured here, but the overall pattern is the same.) Pay particular attention to where the timeline for the Netherlands goes, shown at the top right.

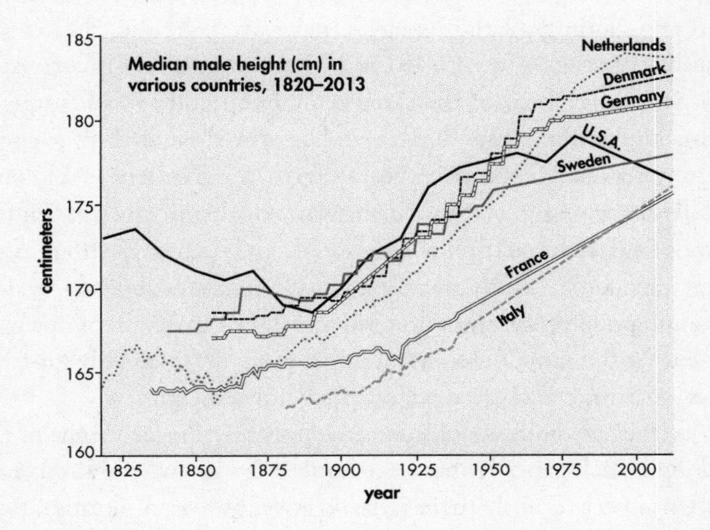

Surprisingly, the Dutch are the tallest men in the world, with an average height of 185 centimeters (about 6 feet 2 inches). There is reportedly a club in Amsterdam for men who are over 6 feet 10, which isn't uncommon. A brief walk down Amsterdam's streets will bring into view both men and women of head-turning height.

This gain in height represents a recent trend, as the chart also shows. There have been steady gains in many countries since 1820, yet the Dutch stand out because they were the shortest Europeans back then. Examination of skeletons in graves from 1850 indicate that Dutch men on average stood around 5 feet 5 inches and women 5 feet 1 inch. (The second-tallest men in 2013, the Danes, were about 6 centimeters [2.3 inches] taller than the Dutch in 1829 and now have fallen slightly behind.) What happened to cause such a dramatic growth spurt in such a short period of time?

Looking for an explanation, Olson consulted other statistics, which revealed that as income went up and the Dutch became more prosperous, wealth was more evenly spread out. Instead of the privileged few gaining all the money, almost everyone did. This more equal distribution of wealth led to a better diet, which is correlated with growing taller. But the same economic trend spread throughout most of Europe, so it doesn't explain why the Dutch in particular grew so tall. To deepen the mystery, city dwellers in Holland actually decreased in stature during parts of the nineteenth century as compared with the rural population. Living in a city, with its high infant mortality, communicable diseases, an impoverished underclass, and polluted air and water, led to a deficit in height of one inch in men. At the same time, urban populations were steadily growing richer, so prosperity isn't a perfect predictor of height.

A clever possibility looks directly to genes. The sequence of the DNA in Dutch genes is about the same as it was two hundred years ago. Until very recently there were no strong waves of immigration, and they wouldn't alter Dutch genes unless there was intermarriage with the newcomers. But what if the reverse were true? It's generally accepted, Olson points out, that our human ancestors were tall. Perhaps the Dutch used to be tall, hundreds of generations ago, but then poor diet caused them to shrink. In that case, a better diet might trigger the ancestral genes, causing a growth spurt.

That's a tenuous possibility, yet any explanation must include genes, especially the epigenome. Since the epigenome is modified according to one's past experiences, what could cause a sudden jump in height? As it happens, one of the best proofs that epigenetics can in a sense *record memories* of past experiences also comes from Holland. The Dutch famine, also known as the *Hongerwinter*, or "hunger winter," has probably taught us more about the effects of epigenetics in humans than any other event. While the Germans were facing the beginnings of defeat in World War II in the extremely harsh winter of 1944, they enforced a food and supplies embargo on the Dutch and began systematically destroying the country's transportation systems and farms. Drastic food shortages resulted, and a famine occurred over the winter of 1944–1945. Food stocks in the cities in western Holland quickly dwindled. The daily adult rations in Amsterdam dropped to below 1,000 calories by the end of November 1944 and to 580 calories by the end of February 1945—only one-quarter the calories needed for health and survival in an adult. The population subsisted mostly on hard bread, small potatoes, sugar, and very little, if any, protein.

Millions of years of evolution have armed us with the ability to survive long periods of malnutrition. The body slows down to conserve energy and resources. Blood pressure and heart rate are decreased, and we begin to live off our own fat. Much of this ability is made possible by changes in the activities of our genes. In some cases, gene activities are turned up and down via epigenetics. The Dutch experience went even deeper, however, showing that DNA changes brought on in adult life can be inherited by the next generations. Studying the children born to survivors of the Dutch famine revealed just this.

Investigators from Harvard obtained the meticulously maintained health and birth records from that time, and as expected, babies born during the famine often had severe health issues. Infants in the womb in the third to ninth months of pregnancy during the famine were born underweight. However, infants in the first trimester toward the end of the *Hongerwinter*—that is, on the cusp before food supplies returned—were actually born larger than average. The differing diets of the mothers created this effect.

The bigger surprises, however, came in studying these offspring after they reached adulthood. Compared with those born outside the famine, adults born during the famine were highly prone to obesity. In fact, there was a doubling of obese individuals among those who were in the womb during the famine, particularly in the second and third trimesters. Some kind of epigenetic memory seems to be at work. We will get to the exact mechanism in a moment.

The Dutch famine studies are important because they opened everyone's eyes to the lifelong effects of prenatal experiences that cause changes in the genome. The beautiful and beloved actress Audrey Hepburn was a child in the Netherlands during the famine. As an adult she suffered from anemia and bouts of clinical depression. She was not alone. Babies who were in the womb during the famine were also more prone to schizophrenia and other psychiatric illnesses. Although not conclusive, some data indicate that when the famine babies had children of their own, the next generation was underweight. Like a conveyor belt, the genome kept passing along a severe food shortage from one generation to the next.

THE CONVEYOR BELT OF EXPERIENCE

This new knowledge about inherited traits grew out of terrible suffering, but it sheds light on why better care of mothers during pregnancy is so critical. But controversy surrounds the findings nevertheless. Can the conveyor belt really cross the generation gap? In 2014, data coming from high-quality studies in mice provided the first compelling evidence that transgenerational inheritance can occur in mammals. A Cambridge University geneticist in England, Anne Ferguson-Smith, published findings in the prestigious journal *Science* after testing the epigenetic implications of the Dutch famine in mice. "I decided it was time I actually did some experiments on this myself," she was quoted as saying, "rather than criticize people."

Heated criticism revolved around the key finding that a pregnant mother's diet has a long-lasting impact on the health of her offspring late into their lives. To a strict Darwinian, at the moment the father's sperm fertilizes the mother's egg, the fate of the genes is firmly established in the baby. Ferguson-Smith and her colleagues sought direct evidence by using a strain of mice that could survive on an extremely low-calorie diet. As expected, the mice had extremely underweight offspring that were later prone to diabetes. The males in this litter then fathered another generation, and the second generation of mice also went on to get diabetes even though they consumed a normal diet. These startling findings provided evidence that the genetic conveyor belt is real.

The new paradigm opens enormous vistas. Pregnant mothers are already advised not to smoke or drink alcohol during pregnancy. Exposing the fetus to toxins raises the risk of birth defects. It's good to heed the statistics about risks. But what about enhancing a baby in the womb? You've probably read stories about pregnant mothers who play Mozart to their babies in the womb, and other reports about how a fetus in utero can be affected by stressful situations that the mother-to-be undergoes. A major theme of this book is to give your genes the lifestyle with which they can optimally function. This would be doubly true if you are deciding the genetic inheritance of one, two, or more generations in the future. And what if the conveyor belt were loaded with such optimal experiences that children and grandchildren were given the best possible start in life by "soft" inheritance? To us, this is far more inspiring than schemes for manipulating the genome of embryos with the aim of a genetically "perfect" baby. The science of transformation doesn't always have to mean implants and syringes.

To bring about a generation of children with the best traits that can be passed on through soft inheritance, we must look deeper into the science behind what this means. To explain how an experience leads to genetic changes we need a new term: *epigenetic marks*. Such marks are the fingerprints of change. They are key for solving the mystery of how any lifestyle change influences our genes, not just a drastic change like the "hunger winter." Epigenetic events can also program DNA by chemical modifications of the pillow-like proteins (called histones) that surround and cushion the DNA. These cushions also decide what stretch of the DNA making up a gene is exposed to other proteins that turn the gene on and off, up and down in activity, and even what type of proteins or RNA the gene will manufacture.

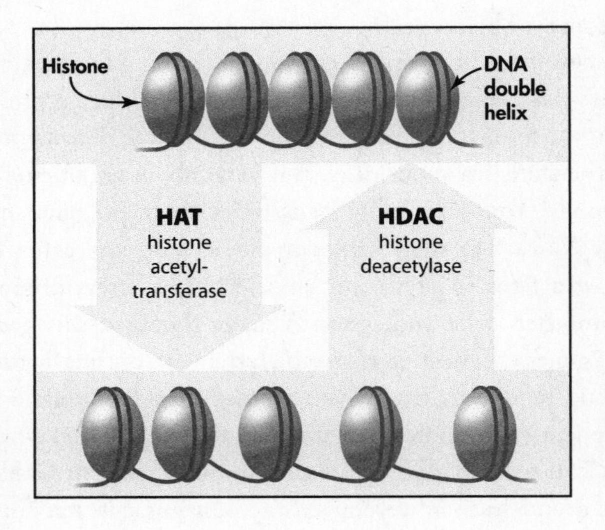

Imagine, then, that the body has begun to be food deprived and eventually begins to starve. How does a pregnant mother's body respond? We can observe it wasting away, but invisibly her epigenome is creating genetic alterations. The cushion-like proteins that surround the DNA start to interact differently with the DNA, leaving epigenetic marks. The marks can be of various kinds, involving specific enzymes with names like methylase and histone deacetylases (HDACs). Even tiny bits of RNA (micro-RNAs) can do the job. You don't need to remember how the chemistry of epigenetic programming works, but a growing body of evidence indicates that diet, behavior, stress levels, and chemical pollutants can all affect gene activity and thus one's survival and well-being.*

The epigenetic marks that are probably the most studied are those involving "DNA methylation." Wherever there are multiple C bases next to G bases in a chromosome's DNA sequence, there is an increased chance of methylation. If these areas become overly tagged by methylation, gene activity can be turned off.

Methyl marks offer a wide range of clues. For example, many allergies begin early on in fetal development. If an expectant mother eats a diet rich in foods that tag DNA with methyl marks, it's possible that allergies will more likely arise in the child. This means that the same embryo gestating in two different mothers can lead to two different babies despite having identical DNA. One study showed that simply by counting up the methylation marks on the genome of DNA from saliva, researchers can predict someone's age to within five years. The more marks, the older the person is, like reading tread wear on rubber tires. This implies that excessive methylation may be the cause of premature aging and degenerative diseases among the elderly.

Overfeeding mice just after birth has been shown to lead to an excess amount of methyl marks on specific genes that then predispose them to obesity. It is difficult to extrapolate how these effects

*Note: To simplify the highly complex subject of genetic switches, we've focused on methyl marks, but switching also involves other chemical processes, such as acetylation, that we're skipping over. The histone "pillows" are also involved in turning genes off and on or even changing how tightly the DNA helix is wound or folded. Both methylation and acetylation can modify histones and how they bind to the DNA, thereby affecting the activities of genes in the region.

in mice play out in humans. But the Dutch famine, and the experiments it led to, offers some stark testimony.

A CLOUDY ANSWER

But what about the Dutch becoming the tallest men in the world? Sometimes answering a question requires ruling out the false answers first. In this case, we know that no single gene for height is involved, because no such gene exists. If a pregnant woman wants a prediction about how tall her baby will grow up to be, our current understanding of genetics can't tell her. More than twenty genes have been identified that contribute to the height of a child, and their interactions are too complex and slippery for making any accurate predictions.

Even if this side of the story could be sorted out, there are environmental factors that by most estimates account for at least half of the final outcome. These factors include the mother's diet as well as the baby's, but also intangibles like the behavior and lifestyle of the mother and the family environment in which the child is raised. In North Korea and Guatemala, for example, there is chronic malnutrition, and as a result, children grow up stunted. Poor medical care can have the same result, while better general health makes a population grow taller. But the Dutch aren't markedly different in these areas than the rest of Europe. As previously mentioned, over the last two hundred years, despite periods of declining diet, such as in Germany after World War I, a better diet and greater prosperity have led to increased height in all European countries.

What other answers can be ruled out? No new genes entered the Dutch gene pool in sufficient numbers to provide an answer. Even if new genes did mix with the old, there's no evidence that the Dutch started marrying extremely tall newcomers. Nor will survival of the fittest tell the tale, because shorter Dutch men didn't die out after being defeated for food and water by taller men.

However, mating habits could play a part. When the Chinese imperial court began to favor lapdogs, the Pekingese breed emerged by design, beginning with the original dogs from western China more than two thousand years ago. Ancient court documents specify what the ideal Pekingese should look like, the model being a miniature lion. Breeders were instructed to develop a dog with a flat face, large glistening eyes, a mane, short legs, and very small size. In the mind's eye of a Chinese court lady, those qualities were lion-like. To arrive at the ideal, dog breeders kept selecting the smallest pups in a litter and mating them to arrive at smaller dogs. In the same way, other aspects could be encouraged in the breed.

Human beings don't mate by following a breeder's chart, and historically almost everyone got married, so specific traits weren't weeded out, certainly not intentionally. But we do choose our mates consciously, following personal inclinations. If the Dutch came to admire height, and the tallest people were attracted to other tall people, this sequence would produce taller offspring over time. Generally, genetic traits don't favor extremes but return to the mean. There have been humans as short as two feet and as tall as eight feet. But the overwhelming odds are that a baby will be born much closer to the average, growing up to be somewhere between five and six feet tall.

Regression toward the mean, as statisticians call it, also explains why two parents with high IQs can't be guaranteed to produce a child with a high IQ. The genetic component of intelligence (which remains a controversial subject) favors average intelligence, average height, average weight, and so on. Thus it would take generations of Dutch people, a large majority marrying for height, to produce a trend in the population. Once again, the story of inheritance is too complicated for one factor to suffice.

So what now? Once you eliminate the false answers, a new kind of thinking starts to emerge. Dutch men grew taller, not by any simple cause and effect, but because of a cloud or fog of causes. Genes, havior, diet, and various outside influences all played true for all babies, so it must be true for Dutch babies span of two centuries. Out of this cloud of causes, however, we can extract some positive conclusions:

Many factors in the cloud of causes are under our control.

Very few of the causes are deterministic. We are rarely puppets being controlled by our genes.

The cloud of causes is highly adaptable to change.

These are very important conclusions. A cloud changes shape when the wind changes, the temperature goes up and down, weather fronts move, and humidity rises and falls. At any given moment, the clouds you see floating overhead are responding not simply to one of these influences, but to several or all of them. Trying to analyze one at a time isn't valid and sometimes can't be done. It's like trying to predict what the temperature in your house will be if there were five thermostats, each with its own setting for a single zone.

Even under the worst conditions, like the terrible stresses of war, the human genome can find an advantage. During the World War II food shortages in Holland, hospitals noticed an improvement in children hospitalized with the rare intestinal disorder celiac disease. The cause of the disorder was as yet unknown, although it had been hypothesized that diet, and specifically wheat, was involved. A Dutch pediatrician, Dr. Willem Dicke, investigated this connection. When the sick children had almost no bread, they recovered. When the first supplies of bread were allocated to sick children in hospitals, the celiac patients relapsed. This occurrence proved the connection between celiac disease and wheat for the first time. It's now known that celiac disease is an autoimmune disorder with a genetic predisposition that causes an allergic reaction to a gluten protein (gliadin) found in wheat. Similar gluten proteins found in other grains also create this reaction.

Similarly, in countries like Holland and Belgium, where the diet was rich in butter and cheese, the war caused a marked decrease in heart disease, which was attributed to a sudden drop in daily calories and a drastic shortage of butter, milk, and cheese when these countries were occupied by the Nazis. Decades later, losing weight and drastically cutting back on daily fat intake became part of heart-healthy programs for actually reversing heart disease.

A cloud isn't a very satisfying model for doing science, and it's totally inadequate for reaching results in medicine. Doctors are wedded to the linear model of cause and effect. Cause A leads to disorder B, for which the doctor prescribes drug C. But what if the cloud model is actually correct and inescapable? Nobody's living room has five thermostats running independently, but we all have bodies with multiple clocks, biorhythms, and genetic schedules. For this reason, no two people are exactly matched for the day they lost their first baby tooth, entered puberty, felt the first twinge of arthritis, or a myriad of other things that are timed individually. Everything about us moves on a sliding scale.

The question arises, then: How does the human body manage to be so precisely regulated that it synchronizes all its clocks down to the last molecules of hormones, peptides, enzymes, proteins, and so forth? Like a cloud, we are pushed from all directions, but unlike a cloud, our bodies are miracles of complexity that maintain an astonishing amount of control.

Now that we have the complete DNA sequence of the human genome, it is much easier to find genes and mutations associated with the risk for disease. Thousands of disease-associated genes and mutations have been found for disorders ranging from cancer to diabetes, from heart disease to the degenerative brain diseases of old age. Rudy has found several genes and mutations that cause or affect one's risk for Alzheimer's disease (including the first such gene) as well as other insidious neurological disorders like Wilson's disease, a rare condition in which copper accumulates in the cells, leading to serious neurological, psychiatric, and other conditions. As increasing numbers of disease-causing genes have been elucidated, we have learned that roughly 5 percent of disease mutations guarantee disease onset, while the vast majority serve only to increase someone's susceptibility, in concert with the environment and aspects of the person's lifestyle. The bottom line is that humans are a bundle of complex traits for which direct genetic causes aren't yet found and probably won't be found. A more realistic viewpoint of how common diseases are inherited would have DNA serving as the initial blueprint for a building that will be remodeled and repurposed over and over as needed.

Some still believe that knowing what every gene does should be enough to understand all disease, and that when such links are validated, there's the promise that medical therapies to cure genetically linked diseases will follow. But there's a reason why that step hasn't occurred except for a tiny fraction of diseases. You can't figure out what a gene is doing unless you know how it gets switched on and off, up and down, and tweaked to make certain varieties of proteins. No matter how perfect the circuitry for a computer is laid out, the computer is dead until it's turned on. The same holds true for DNA. The triggering mechanism for genes was a mystery that opened the way for the present genetics revolution.

MAKING BETTER MEMORIES

The greatest achievement in Earth's 2.8 billion years of evolution isn't human DNA or even the emergence of life from lifeless molecules swirling in steaming pools of chemical-rich water around the fissures of geysers. Evolution's greatest triumph is memory. Memory is what made life possible. This is clear enough. The antibodies in your immune system contain the memory of all the diseases confronted by the human race. A newborn baby fends off disease by relying on the immune system of its mother, which it has borrowed. Soon the baby's own immunities will develop as the thymus gland, the repository of past battles against invading bacteria and viruses, starts to produce antibodies. The thymus gland expands as it reaches full functioning during adolescence and then shrinks as its task is completed around age twenty-one.

If we focus just on this one process, the role of memory runs deep. The genes of your family line determine which antibodies will appear in you. That's just a twig on the branch of human evolution; that branch leads back to the trunk of the tree, which contains the memory of how to make antibodies in the first place. The roots of the tree are DNA's ability to remember experiences and to encode them for future generations. So the next time you don't catch the cold that is going around, you owe your immunity to the first molecule of DNA.

Epigenetics suggests that our cells can in a sense "remember" everything we have experienced. But a suggestion isn't proof. There's a big difference between remembering your tenth birthday party and a geneticist examining the genetic modifications that encode the memory. Imagine that you are a telegraph operator from decades ago as streams of dots and dashes come across the wire. You can hold the code in your hands and count all the punches in the paper tape, but if you don't know English, the messages are unreadable. In present-day genetics, the code is in our hands, but they're in a language infinitely more difficult than English, the very language of all human experiences.

It's a terrible fate to be at the mercy of your memories, but that's the situation almost everyone finds themselves in. Old fears, wounds, traumatic events, and accidents litter the mind, roaming at will and distorting how we view the present. If you are an agoraphobic, with a fear of open or public places, you can't leave the house without suffering from anxiety. Your fear has made you a slave to memory. In small and large ways we are all enslaved by events that are dead and gone. To be fully alive, you must learn to use your memories, not the other way around.

FEAR AND THE ZAPPED COWS

This is a slightly uncomfortable exercise, but sit for a minute and let a bad memory return. It can be anything—the content doesn't matter. Don't reach for a freshly painful memory. Instead, go back to something that happened when you were a small child. It could be falling off a swing or getting separated from your mother in the grocery store. What do you notice? First, that the memory exists; second, that you can retrieve it. Depending on how deep the memory goes, you will also notice that it feels like real life repeating itself.

The same part of the visual cortex that sees a train wreck or a battle scene comes into play when a person visualizes the wreck or the battle by recalling it.

Everything you are noticing is reflected in your epigenome. Let's go a step further. When the children of the Dutch famine became vulnerable to obesity, diabetes, and heart disease, those memories could be traced to their mothers' experience of near starvation. The children couldn't see this experience in their mind's eye, and yet they inherited a molecular memory nonetheless. A striking study published in the high-impact journal *Nature Neuroscience* in 2014 added new evidence about memory's effect on DNA, only in this case the driver was not diet, but fear. In this study, scientists trained mice to fear the scent of the chemical acetophenone (which is pleasant, like orange blossoms and cherries) by giving them a mild electric shock whenever the smell was introduced.

The shocks produced a stress reaction in the mice, which could be observed in their nervous, shuddering behavior. After a while it wasn't necessary to deliver the shocks. Merely the smell of acetophenone was enough to produce the stress reaction. A maker of horror movies can do much the same thing by showing a dark room with the sound of a creaking door. The heroine's eyes dart in fear, and what's happening in the audience? These harmless images and sounds produce the anticipation of something horrible about to happen. Evidence of the stress response will appear in most viewers.

But the mouse study that associated a harmless odor with an electrical shock went further. This acquired fear in adulthood was inherited by the mice's offspring and even by the next generation. The children and grandchildren of the fear-conditioned mice had never experienced the fragrance of the acetophenone before, but they shuddered as soon as they smelled it, simply because their parents were conditioned to associate the odor with pain. The researchers then looked at the gene that makes the protein receptor needed to smell the chemical and found that it had been epigenetically modified by methylation.

Folk wisdom has known about this phenomenon forever, as expressed in a piece of cracker-barrel wisdom from Mark Twain: "If a cat sits on a hot stove, that cat won't sit on a hot stove again. That cat won't sit on a cold stove either." In the same vein, the wisdom behind getting back on a horse after you fall off is based on the instinctive knowledge that fear can make a lasting impression unless you counter it as soon as possible. Of course, this type of conditioning is mediated by memories maintained by the neural networks in your brain. The same experiences can chemically modify your genome to create a parallel "molecular memory."

We've repeated several times that DNA is responsible for both stability and change. Now we've arrived at a new wrinkle. How do our brain and genes determine the difference between real danger (a hot stove) and imagined danger (a cold stove)? Animals apparently don't, as proved by studies of cattle trained by electric fences. The first step is to enclose cattle in a tight corral bounded by an electric fence that delivers a harmless shock if touched. The electrical current runs through one thin wire.

After only a day, and sometimes just an hour, the zapped cows have learned to avoid the fence. They can then be released into a grazing area that is fenced in by a single wire. Even though the cattle could easily break through this barrier, training them with an electrified wire keeps then inside. Thus the old principle of physically hemming the cows in with barriers like rail fences is exchanged for a psychological barrier. It's difficult for old ranchers to accept that a psychological fence can be more powerful than a physical one, but in experiments in which hungry cows were separated from a bale of hay by a single strand of electric wire, they would not break through it to get at the food.

Is this form of psychological training inheritable? So it appears, as evidenced once again by cattle. To keep cattle from wandering down a road, ranchers install grids, usually steel railing, with gaps between the rails. Yet it appears that actual cattle grids aren't necessary. They can be fooled by phony grids, as described by Rupert Sheldrake, a British biologist famed for adventurous thought and investigation. (This trait has made him a groundbreaking thinker, an audacious rebel, an outlier from mainstream biology, or someone far too credulous about mysterious phenomena, depending on the view of him you take. We greatly appreciate his daring.) In a *New Scientist* article from 1988, Sheldrake writes:

> Ranchers throughout the American West have found that they can save money on cattle grids by using fake grids instead, consisting of stripes painted across the road. . . . Real cattle grids make it physically impossible for cattle to walk across them. However, cattle do not usually try to cross them; they avoid them. The illusory grids work just like real ones. When cattle approach them, they "put on brakes with all four feet," as one rancher expressed it to me.

Although Sheldrake picked up on this phenomenon from American friends he was visiting in Nevada, the implications resonated with him. For decades Sheldrake had been almost a lone voice proposing that memories can be passed down from one generation to the next. Undaunted by ridicule from orthodox geneticists—this was long before the advent of epigenetics—he wrote farseeing books like *A New Science of Life* (1981) and *The Presence of the Past* (1988) to amass the mounting evidence that inheritance across the generations was real. These are still among the most fascinating and eye-opening books on the subject of memory as the major force in evolution. As Sheldrake explains:

> According to my hypothesis . . . organisms inherit habits from previous members of their species. This collective

> memory, I suggest, is inherent in fields, called morphic fields, and is transmitted through both time and space. . . . From this point of view, cattle confronted for the first time by grids, or by things that look like grids, would tend to avoid them because of [inheritance] from other cattle that had learnt by experience not to try to cross them.

A skeptic would protest that other, more conventional explanations must be at work. It could be that cows don't inherit an avoidance of cattle grids but acquire it individually through painful exposure to real grids, or else they somehow pick it up from more experienced members of the herd.

Sheldrake responds:

> This does not seem to be the case. Ranchers have told me that herds not previously exposed to real cattle grids will avoid the fake ones. This has also been found by researchers in the departments of animal science at Colorado State University and Texas Agricultural and Mechanical University, with whom I have been in correspondence. Ted Friend, of Texas A&M, has tested the response of several hundred head of cattle to painted grids, and has found that naïve animals avoid them just as much as those previously exposed to real grids.

Is this also a possibility among humans? Inheriting a behavioral trait might explain why Mohawk Indians have worked for generations on the construction of New York skyscrapers—they walk on the beams hundreds of feet in the air, apparently with no fear of falling. Did they inherit this trait? Does the same kind of inheritance account for why Russian chess players have won the world championship many times over?

Yet the effect of memory inherited across generations is soft enough that it can be reversed, at least in animals. Writing about the cattle who shy away from phony grids, Sheldrake says:

> Nevertheless, the spell of a fake grid can be broken. If cows are driven towards one under pressure, or if food is placed on the other side, a few will jump it; but sometimes one will examine it closely and then simply walk across. If one member of a herd does this, then the others soon follow. Thereafter, the phony grid ceases to act as a barrier.

At least some sheep and horses also show an innate aversion to crossing painted grids. By contrast, in perhaps the only experiment of this kind ever carried out with pigs, the animals ran up to the painted grid, sniffed it, and started to lick it up. The Texas researchers had used a washable water-based paint with a flour-and-egg base.

Noticing these aspects of memory comes easily. We are all expert time travelers in our minds. But as skillful as we are at storing a memory and recalling it, we are much worse at erasing bad memories. Memories are sticky. Years of therapy can fail to undo the power of old traumas. Drugs and alcohol only mask them temporarily. Denial pushes a bad memory under the carpet, but there's no guarantee that it will stay there.

Genetics tells us that any past experience, good or bad, is sticky because it has taken its place, using chemical bonds, deep inside the cell, in the nucleus where DNA resides. In a molecule of salt, atoms of sodium and chlorine are tightly bound together. A lot depends on their remaining stuck, because if you poured out some salt and it separated into its components, the release of chlorine gas would be poisonous. Likewise, it's necessary for DNA's bonds to remain secure or life would vanish into a cloud of atoms.

Life is about the persistence of memory. Until recently, the only memories available to geneticists were the rungs that connect the double helix of DNA, and these were fixed in place long, long ago in evolutionary time. However, epigenetics now uses chemistry to create genetic memories of past experiences, which are much more recent and intimate than the 2.8-billion-year-old memories that originally built the DNA molecule.

FROM ADAPTATION TO TRANSFORMATION

Genetics is well under way with its present revolution, but how does it impact you in your daily life? Simply, through adaptation. Dinosaurs adjusted so well to their environment that they dominated life on Earth as the major predator. They pushed the climate barrier, moving into colder zones that are now in the Arctic (because of the shift of tectonic plates). In their diets, some dinosaurs were vegetarians and some carnivores. But superb as this adaptability was, a cataclysmic event destroyed the dinosaurs. A giant meteor collision with Earth, thought to be in the region of present-day Yucatán in Mexico, created an overnight change in climate. Dust from the impact clouded the sun all over the globe, the temperature dropped precipitously, and dinosaur DNA didn't have time to change.

Or did it? Some present-day reptiles survive freezing climates by hibernating through the winter, which allows snakes to live in New England, for example. But adaptation takes a long time, eons even, only if a species has to wait for random mutations. Adaptation can occur much more quickly in an individual through gene expression.

THE GOAT WHO WOULD BE HUMAN

In 1942 a Dutch veterinarian and anatomist named E. J. Slijper reported on a goat born in the 1920s with no functional forelegs. The baby goat adapted to its unfortunate condition by learning how to hop, kangaroo-like, on its hind legs. The goat survived a year before dying accidentally. When Slijper performed an autopsy, he found several surprises. The goat's hind leg bones had elongated. Its spine was S shaped, like our human spine, and the bones were attached to the muscles in a way that looked more like those of a human than those of a goat. Two other human characteristics had started to form—a broader, thicker plate of bone protecting the knee and a rounded inner cavity in the abdomen.

It's startling to think that in one year a new behavior, walking upright, could make it appear as if a goat were becoming human, or at least like an animal that walks on two legs, because all of these changes are associated with the evolution of bipedal motion. Gene activities had changed to remodel the goat's anatomy. For a long time Slijper's goat attracted no serious attention. In the standard Darwinian view, how humans learned to walk on two legs was by random mutations that changed our carriage from the stooped posture of other primates, and that such mutations almost always occur one at a time. Even without Slijper's observations, it's quite challenging for evolutionists to explain how all the anatomical adjustments needed for human beings to walk upright could credibly occur one at a time. However, they all work together, and the goat proved that they could arise together, not as mutations but as adaptations. Can the epigenome actually pass along a complete and interconnected set of changes?

While the argument rages back and forth, there is no backing down from the speed of adaptations in human beings. The question of how much your lifestyle will affect your children and grandchildren hasn't been settled. But the changes occurring in you are indisputable.

This is why identical twins are not actually identical. Beginning at birth, they start to live different lives and thus become different people, despite carrying virtually duplicate genomes. Identical twins can be quite variable in their susceptibility to disease and in their behavior. Genetic studies of identical twins have traditionally been used to determine what is referred to as the heritability of disease. If one twin gets a certain disease, what are the odds that the other will get that disease within fifteen years or so? It's a simple calculation, actually. After studying hundreds of pairs of identical twins, researchers determined that the probability for Alzheimer's disease occurring in both twins is 79 percent if one of them is afflicted. This means that lifestyle accounts for 21 percent of the probability of developing Alzheimer's, even with identical genomes.

In contrast, for Parkinson's disease, the heritability is only about 5 percent; therefore lifestyle would appear to play a hugely greater role. For hip fractures under the age of seventy, the heritability is 68 percent, but after age seventy it goes down to 47 percent. For coronary artery disease, the inheritability is about 50 percent, no more than random chance. For various cancers—colon, prostate, breast, and lung—the heritability in identical twins ranges from 25 percent to 40 percent, which is why the current view holds that the majority of cancers, perhaps a large majority, are preventable. Epigenetic changes associated with cancer can be induced by factors like chronic exposure to asbestos, solvents, and cigarette smoke. Yet these cancer-causing epigenetic changes could be offset with a healthy diet and exercise—it's a highly promising possibility.

CHANGE IS IN THE AIR

Physical changes don't always need physical causes. Sometimes the stimulus can simply be a word. If you meet someone new and fall in love, there's a dramatic shift in brain activity—this has been thoroughly documented—and if the person you're smitten with says "I love you," as opposed to "I'm seeing someone else," the gene expression in the emotional center of your brain will be dramatically altered. At the same time, chemical messages sent via the endocrine system will create an adaptation in your heart and other organs. To be accepted by a beloved can make you lovesick; to be rejected makes you heartsick. There's a unique gene expression for both.

There's solid science behind these age-old experiences. In a 1991 study by microbiologists at the University of Alabama, mice were injected with a chemical that boosted their immune system. This chemical, known as poly I:C (polyinosinic:polycytidylic acid), causes greater activity in part of the immune system called natural killer cells. At the same time as the mice received the poly I:C injection, the odor of camphor was released in the air. The mice were quickly trained to associate the two, and thereafter a tiny amount of poly I:C would be enough to stimulate the mice's natural killer cells as long as the smell of camphor was in the air.

The mice's bodies on their own manufactured the chemicals needed to stimulate their immune system. All they needed was a small trigger. This is an impressive finding, because it shows that genes can adapt in a specific direction with very little motivation. The actual molecules of camphor passing from the nose to the brain of a mouse have no effect on the immune system. It was the *association* of the camphor that created the effect. We've gone one step further than the zapped cattle, whose behavior was changed by remembering the pain of being shocked. The mice did no conscious learning. Their bodies adapted without the mind (such as it is) having to learn or even think.

Human beings *can* think, of course, but our body is constantly being affected when we aren't aware of it. As far as smell is concerned, pheromones given off by the skin are connected to sexual attraction in mammals and seem to play a part in human attraction. In an experiment to test aromatherapy, researchers found that people reliably report a positive mood change after smelling lemon oil as compared with no change after smelling lavender or odorless water. This mood elevation happened whether the subjects had ever experienced aromatherapy before. In fact, one group wasn't told anything about the aromas or what to expect, and their mood also improved upon smelling the lemon oil.

Yet the power of expectation is undeniably strong. In the placebo effect, a subject is given an inert sugar pill and told that it's a drug for relieving symptoms like pain or nausea, and in 30 to 50 percent of people, the body steps in and produces the chemicals needed to bring about the expected result. As familiar as the placebo effect has become, it's still remarkable that mere words ("This will help your nausea") can trigger such a specific response in the connection between brain and stomach. You can even give the subject a drug that *causes* nausea, and just by being told that it's an anti-nausea pill, some people will experience that their nausea goes away. To complete the picture, there's the nocebo effect, in which giving someone a harmless sugar pill and telling them that they won't feel any benefit from it can even create negative effects.

We seem to have wandered rather far from how adaptation failed the dinosaurs, but all of these findings are highly relevant. If a mere odor or the words "This will make you feel better" can alter gene expression, and if a totally inert substance can create nausea or make it go away, the whole world of adaptation is wide open. Instead of being like Pavlov's dogs, which salivated every time they heard a bell associated with mealtime, humans insert another step—interpretation.

In a mouse trained to associate camphor with a stronger immune response, there isn't any interpretation. Stimulus leads to response. But all attempts to train human behavior run at least a fifty-fifty chance of failing. Positive incentives like money, power, and pleasure affect everyone, but there's always the person who says no and walks away. Negative incentives like physical punishment, bullying, and extortion are very likely to make people do what their tormentors want them to do, but there are always some who resist and don't comply. Between the stimulus and the response comes the conscious mind and its ability to interpret the situation and respond accordingly.

So what we have is a feedback loop that is at work in every experience. There's a triggering event A, leading to mental interpretation B, resulting in response C. This response is remembered by the mind, and the next time the same event A arises, the response won't be exactly the same. This feedback loop is like a never-ending conversation between mind, body, and the outside world. We adapt quickly and constantly.

This result became even more fascinating when the experiments took the same odor of camphor and introduced it while the mice were injected with a drug that lowered immune response. Once more, after a period of time it took only the camphor smell to impair the mice's immune response. In other words, the same stimulus (camphor) could induce a specific response and its exact opposite.

ADAPT FIRST, MUTATE LATER

In spite of the growing base of evidence supporting epigenetics, some evolutionary biologists are certain to insist that the evolution of our species is entirely random and based solely on natural selection. To even imply that there may be some highly interactive epigenetic program driving the evolution of our species will prompt many a staunch, card-carrying evolutionary biologist to froth at the mouth and label you a "creationist" touting notions of "intelligent design." We are certainly not suggesting "intelligent design." However, considering the mounting evidence for the effects of epigenetics on overall health, it's time to seriously consider what the new genetics is teaching us about our own evolution.

Current findings could make a life-or-death difference. For nearly three decades at Ohio State University, Professor Janice

Kiecolt-Glaser and her colleagues have been examining the effects of chronic stress on the immune system. The general picture was already well known. If you are subjected to repeated stress, resistance to disease goes down. In addition, you run the risk of developing disorders like heart disease and hypertension. But people are much less familiar with the dangers of everyday stress, the kind we don't like but feel we should put up with.

Kiecolt-Glaser's group looked at a stress that has become far more common recently, taking care of someone with Alzheimer's disease. The baby boom generation is being sidelined more and more by taking on the responsibility for aging parents with Alzheimer's, and because professional care is limited and too expensive, millions of grown children find themselves being the last resort for caregiving. As much as we love our parents, round-the-clock caregiving imposes serious chronic stress, day in and day out.

A genetic price is being paid. As a research website from Ohio State reported: "Earlier work by other researchers has shown that mothers caring for chronically ill children developed changes in their chromosomes that effectively amounted to several years of additional aging among those caregivers." When their attention turned specifically to Alzheimer's caregivers, it wasn't surprising that Kiecolt-Glaser's team found higher indications of depression and other psychological effects. But they also wanted to target the specific cells that showed evidence of genetic changes.

They found them in the telomeres of immune cells. Telomeres, you recall, are the caps that end a DNA sequence like the period at the end of a sentence. Telomeres fray as cells divide over and over, which gives a marker for aging. "We believe that the changes in these immune cells represent the whole cell population in the body, suggesting that all the body's cells have aged that same amount," says Kiecolt-Glaser. She estimates that this accelerated aging deprives Alzheimer's caregivers of four to eight years of life. In other words, the adaptability of our bodies has serious limitations.

Kiecolt-Glaser pointed out that there are ample existing data showing that stressed caregivers die sooner than people not in that role. "Now we have a good biological reason for why this is the case," she said. As Rudy was quick to appreciate while sequencing through entire genomes of over fifteen hundred Alzheimer's patients and their healthy siblings, the genome is chock-full of repetitive sequences of A's, C's, T's, and G's. Some of these repeat sequences in the DNA can bind certain proteins residing deeply inside the nucleus of the cell in order to control the activities of genes in their vicinity. Other repeats lie at the tips of chromosomes, and their length is controlled by proteins such as telomerase. The longer the chromosome tips stay stable (rebuilt by telomerase), the longer the cell survives.

The fact is, over our lifetime we are adapting to our environment every day by modifying our bodies, including at the level of our gene activities. Your next meal, your next mood, your next hour of exercise is modifying your body in an endless flow of change. Darwin explained how a species adapts to the environment over eons of time, allowing for tens of millions of years during which dinosaurs arose and then changed into birds. Flight feathers are a physical adaptation to environmental pressure and nothing more to a strict Darwinist. But in fact our genomes are adapting in real time at every moment of our lives in the form of gene activity. Is it possible that these adaptations are a driving force all on their own?

This is a hot-button issue right now. For the vast majority of evolutionary biologists, putting adaptation before mutation is unacceptable. But there are exceptions. In a *New Scientist* article from January 2015 titled "Adapt First, Mutate Later," reporter Colin Barras brings up Slijper's goat in a new context. A primitive fish from Africa called the bichir has the ability to survive on land. As an adaptation, walking on land aids survival in the drought season by allowing the bichir to leave a dried-up pool to find fresh water as well as new sources of food and a wider territory to colonize. Other species have the same adaptation. When a walking catfish from

Southeast Asia (*Clarias batrachus*) escaped into the wild in Florida, it became highly invasive by traveling over land. The walking catfish doesn't use two legs but wriggles along, propped up on its front or pectoral fins, which keep its head up. As long as they stay moist, these catfish can remain out of the water almost indefinitely.

This adaptation to land travel reminded Emily Standen, an evolutionist at the University of Ottawa, of how ancestral fish emerged from the oceans hundreds of millions of years ago. Recently a 360-million-year-old fossil has caused a sensation by providing physical evidence of this epochal change in life on Earth. A newly discovered fossil fish named *Tiktaalik roseae* had a skeleton that was fishlike but with new features that were like tetrapods, or four-legged land dwellers. Standen specializes in the mechanics of evolving species, and she wondered whether these same adaptations could be sped up—and they could, quite dramatically.

Standen and her team raised bichir fish on land, and being forced to wriggle on their fins more than they normally would in the wild, the fish changed their behavior, becoming more efficient walkers. They placed their fins closer to their bodies and raised their heads up higher. Their skeletons also showed developmental changes—the bones supporting the fins had changed shape in response to higher gravity (fish in water weigh less). Like Slijper's goat, a whole group of necessary adaptations had formed. It will take time to see how far this line of research will take us, but it already suggests exactly what the title of the *New Scientist* article says: "Adapt first, mutate later."

THE RUSSIAN DOLL PROBLEM

This revisionist thinking is a lot to take in, but we assure you that it is all leading to something great. Replacing the simple cause-and-effect model of evolution for a cloud of vague influences is unsettling. The same holds true for your body right this minute. On any given day it is bombarded with influences—through food, behavior, mental activity, the five senses, and everything happening in the environment. Which one will be decisive? Genes can predispose you to depression or type 2 diabetes or certain kinds of cancer, yet only a percentage of people with such a predisposition are going to have the gene activated. Locating the specific factor, or factors, that will activate a specific gene is like throwing a deck of cards in the air and plucking out the ace of spades as they scatter.

Scientists don't like giving up straight-line cause and effect. Many hate the very idea. So we are left with a model that looks like a traditional *matryoshka*, or Russian nesting doll, in which inside the biggest doll is a smaller one, then inside it a still smaller one, and so on, until an extremely tiny last doll. Nesting dolls are delightful, but what if you claimed that the biggest doll was actually *built* by the one inside it, and that one by the next smallest, and so on?

That's essentially where genetics has led us. Sometimes the genetic picture is uncomplicated enough that no ambiguity arises. Imagine you see one white flamingo standing out among thousands of pink ones. What caused it to be white? A linear sequence of reasoning gives the answer. First comes a species, the genus *Phoenicopterus*, which contains six species of flamingos divided between the Americas and Africa. Each has a dominant gene that produces pink feathers generation after generation. But all genes can mutate or fail to appear, leading at random to albinism in a single chick. The number of chicks born with white feathers can be statistically predicted, and there the story ends.

We're using Russian-doll reasoning here, going to smaller and smaller levels of Nature in search of causes. This is the reductionist method, which has time-honored value in science. Chasing nature down to its smallest component is the very business of science, whether it's a physicist chasing down subatomic particles or a geneticist chasing down methyl marks on a gene. But there's a problem here, and it's quite crucial.

Consider someone who has become obese, joining the current

epidemic of obesity that has swept through developed countries. There are many theories about why an individual becomes obese. Stress, hormonal imbalance, bad eating habits from childhood, and the excess of refined sugar and starches in the modern diet have all been suggested. Using Russian-doll reasoning, the eventual explanation would be traced to the genetic level. Although there was once a committed search for "the obesity gene," bolstered by statistical evidence showing that overweight runs in families, that project met with only limited success, identifying some genes (e.g., the FTO gene) that carry DNA variants mildly predisposed to obesity. As with disorders like schizophrenia that have a genetic component, the genetic influence at best provides a predisposition.

Today, a smaller doll has been found in the form of epigenetics and the switches it controls. Almost every factor that might contribute to obesity, whether it's too much stress, excess sugar, bad eating habits, or hormonal imbalance, would theoretically be regulated by the epigenome, the switching station that turns experience into genetic alterations. But here the reductionist line of reasoning hits a wall. It is extremely difficult to tell which particular experience creates which mark on which gene, thereby shifting gene activity. Some people grow obese with or without stress, with or without sugar, and so on. As a result, it is impossible to predict with any accuracy how past or future experiences reliably alter your gene activity. The cloud of causes that surrounds why Dutch men suddenly grew so tall surrounds a great deal of epigenetics. *Something* is creating methyl marks, but the mark is material in nature while the *something* that caused it often is not. An environmental toxin can cause epigenetic changes, but so can a strong emotion, like fear, at least in mice so far.

If you look deeper, the basic assumption that a material cause of epigenetic marks must be at work turns out to be wobbly. It is the entire range of life experience, from physical interactions to emotional reactions, that govern the chemical modification of certain genes with methyl marks. A methyl mark, which you recall is the most studied method for the epigenome to modify a gene, is extremely small. Chemically, a methyl group is tiny, no more than a carbon atom linked to three hydrogen atoms. Methylation marks only the C (cytosine) base pair, sticking to it like a remora, or sucker fish, to the belly of a shark—the cytosine molecule is forty times bigger. It's been shown that when DNA is modified with more methyl marks, some portion of it is switched off. So we seem to be at the smallest doll, the one that switches all the bigger ones. Ninety percent of the modifications in DNA associated with disease are located in switching areas of the gene. Moreover, epigenetics has a remarkable effect on prenatal development, personality and behavioral tics, and susceptibility to disease above and beyond the genes and mutations inherited from our parents.

How your mother lived her life while carrying you in the womb may potentially affect your own gene activities and your risk for disease decades later. Canadian researchers at the University of Lethbridge subjected adult rats to stressful conditions and then studied their offspring. The daughter rats of stressed mothers had shorter pregnancies. Even the granddaughter rats, the mothers of which were not stressed, had shorter pregnancies. The researchers proposed that this occurrence was due to epigenetics. More specifically, they stated that the epigenetic changes brought on by stress involve what are called micro-RNAs,* tiny segments of RNA made from the genome that then regulate gene activity.

Leaving aside potential abnormalities that medical research can focus on, switching is how all of us got here. It's basic to the journey by which a single fertilized cell in a mother's womb grows into a fully formed healthy baby. As this first cell divides, every future cell contains the same DNA. But to develop a baby, there have to

*Note: The DNA between the genes used to be called "junk" DNA. However, we now know that the DNA between genes (or intergenic DNA) can be used to produce tiny molecules called micro-RNAs, which control gene activities throughout the genome.

be liver cells, heart cells, brain cells, and so on, all different from one another. The epigenome and its marks regulate the difference. It was realized that a map of the epigenome was urgently needed in order to locate how each type of cell is determined in the development of an embryo in the womb. Four countries—the United States, France, Germany, and the United Kingdom—have funded the Human Epigenome Project, whose mission is to show where all the relevant marks are, or in official language, to "identify, catalog, and interpret genome-wide DNA methylation patterns of all human genes in all major tissues."

With the participation of over two hundred scientists, a milestone was marked in February 2015 by the publication of twenty-four papers describing, out of the millions of switches involved, those that determine the development of over one hundred types of cells in our bodies. This effort involved thousands of experiments with adult tissue as well as fetal and stem cells. (In theory, counting all the spots on all the leopards in the world would be easier.) The chemicals that regulate different kinds of cells were already known, and sometimes the switches for them aren't close to the affected gene. In fact, switch A can be located at a considerable distance from gene B. In such cases the researchers sometimes had to infer the switch's role by looking at the chemical regulator. If it was present in a cell, they inferred that the switch was turned on.

PARENTS, BABIES, AND GENES

Arriving at this portion of the epigenome map was an exciting development. Switching key genes on and off potentially might be the best route to preventing and curing a host of diseases. As the researchers acknowledge, locating all these switches gives them mountains of new data, but that's only a beginning. In the activity of DNA, switches interact; they form circuits called networks; they can even act on the genes from a distance. Unraveling all the circuitry doesn't indicate why the activity arises, any more than mapping the location of every telephone in a city tells you what people are saying to each other when they call. Different regions of the genome can be turned on in parallel via epigenetics owing to a three-dimensional reorganization of the genome (such as folding the DNA strand into a loop) that brings those regions into close proximity.

There is also the effect that epigenetics has on a child's early life after leaving the womb. This period is like a pivot between the mother's epigenetic influence and the experiences that belong to the infant. How important is the overlap between the two? This question is central to medical issues surrounding infants, and one of these is peanut allergies. As reported in the *New York Times* in February 2015, about 2 percent of children in the United States are allergic to peanuts, a number that has quadrupled since 1997. No one can explain why, but there's been a sharp rise in all allergies in the past few decades, which also remains a mystery. This rise holds true across all Western countries.

A child with a strong peanut allergy can potentially die from exposure to even a small trace of peanuts in food. The standard recommendation has been that giving peanut butter and other peanut-related foods to infants increases their risk for developing the allergy. But a compelling 2014 study published in the *New England Journal of Medicine* has turned conventional wisdom on its head. Feeding infants foods like peanut butter early in life "dramatically decreases the risk of development of peanut allergy," the study's authors concluded. This was heartening news, since it indicated that a step in infant care could reduce or even reverse a rising trend.

The new study was based in London, where 530 infants considered at risk for developing peanut allergy (for example, they might already be allergic to eggs or milk) were divided into two groups. Starting when the infants were between four and eleven months old, one group was fed food containing peanuts, while the other had such foods withheld. By age five, the group exposed to peanuts had far less incidence of allergy, 1.9 percent as compared with 13.7 percent for those whose parents avoided feeding them peanut foods. In fact, it was speculated that having concerned parents keep peanuts away from their infant children might have actually caused the dramatic rise in peanut allergies.

For quite a while parents have been confused over the issue of allergies and newborns, not just relating to peanuts. Before this new finding, the data weren't clear. As we've discussed, a newborn baby inherits its mother's immune system, which serves as a bridge while the baby begins to develop its own antibodies. The thymus gland, located in the chest roughly between the lungs and in front of the heart, is where the immune system's T cells mature. When the body is invaded by outside viruses, bacteria, or everyday substances like pollen, T cells are responsible for recognizing which invaders to repel. An allergy is like a case of mistaken identity, in which an innocent substance is identified as a foe, leading to an allergic reaction created by the body itself, not by the invader.

The thymus is at its most active right after birth up through childhood; once someone has developed a full complement of T cells, the organ atrophies after puberty. The issue with allergies centers on how much of our immunity is inherited genetically and how much is influenced by the environment after we're born. To explain the alarming rise in allergies in developed countries, it would seem that the more polluted the environment, the worse the problem should be. But after the fall of the Soviet Union in 1991 and the opening up of its satellite countries, which generally have much higher pollution rates than the United States or Western Europe, investigators were stymied to find that highly polluted areas in Eastern Europe showed lower allergy rates than in the West.

Then it was thought that the reverse is true: Western countries are overly clean and sanitized, depriving the immune system of exposure to substances that it needs to adapt to. Therefore the peanut-allergy finding could be very significant. American Academy of

Pediatrics guidelines issued in 2000 recommended that infants up to age three shouldn't eat foods with peanuts in them if they were at risk for developing the allergy. By 2008 the academy acknowledged that there was no conclusive evidence that avoiding peanuts was effective beyond the age of four to six months. But there was as yet no study showing that it was correct to stop avoiding peanuts at all. The first real clue came in a 2008 survey published in the *Journal of Allergy and Clinical Immunology,* which found that the number of children with peanut allergy in Israel was one-tenth that of Jewish children in the United Kingdom. The significant difference seemed to be that Israeli children consume peanut foods in their first year, especially Bamba, a popular snack that combines puffed corn and peanut butter, while British children don't if their parents are allergy conscious.

The new study, however, doesn't apply to other foods that children develop allergies to. And two major questions remain to be answered: First, if the children who were fed peanut foods stop eating them, are they liable to develop the allergy? This question is being studied in a follow-up with the original subjects. Second, are the results applicable to kids at low risk for food allergies? That's unknown, but researchers tend to feel that eating peanut food will do them no harm. Asking anxious parents to change their habits, however, may be difficult, since standard care has made such an issue of avoiding the "wrong" foods.

We've gone into some detail, not because we have the answer to allergies, but to make it clear how uncertain environmental influences can be, even though it's known in a general way that epigenetic marks are sensitive to them. The miraculous development of a human from embryo to infant, toddler, adolescent, and adult involves an intricate dance between genes and the environment. In mammals, interactions between the newborn child and its parents can have profound effects on the child's health decades later. Although many findings in this area have emerged only from mouse and rat studies, there is increasing evidence that they may pertain to humans as well. For example, mounting evidence shows that early-life abuse, neglect, and mistreatment lead to epigenetic effects on gene activity that adversely affect physical and mental health later in life.

For good or ill, early events shaping the bonds between parent and child have profound effects on the child's brain development and personality. But how do these bonds get established? Increasingly, studies show that epigenetic modifications of the child's genes are largely responsible, guided by childhood experiences that begin with the earliest days of life. When a mother acts detached from her child, there can be a dysfunctional hypothalamic-pituitary-adrenal (HPA) response associated with stress, impaired cognitive development, and the elevation of toxic cortisol, as measured in the child's saliva.

Some abused children die young, and in these tragic cases their brains can be studied at autopsy. Research of this type has shown clear evidence of epigenetic modification (increased methylation) of the gene NR3C1, which results in nerve cell death in the brain region known as the hippocampus, used for short-term memory. In living children, the same gene modification can be found in the saliva of emotionally, physically, and sexually abused kids. Such damage can lead to subsequent psychopathic behavior.

These findings extend the long-held understanding that early abuse and neglect have profound psychological effects. Now we can trace the damage to the cellular level. In the search for the biological changes that underlie these events, epigenetic pathways controlling gene expression in the brain are increasingly being implicated. By the same token, it may be possible in the future to test the effectiveness of psychological or drug therapies by looking to see if the ill effects in the epigenome have been reversed.

Progress has already been made in animal trials. In 2004 a study at McGill University conducted by neuroscientist Dr. Michael

Meaney showed that baby rats who were groomed (licked) often by their mothers had increased levels of glucocorticoid receptors in the brain, resulting in a reduction in anxiety and aggressive behavior. How were these behavioral changes achieved? Again, by epigenetics. Mice who received affectionate nurturing and grooming by their mothers underwent less modification of their glucocorticoid receptor genes by methylation, resulting in decreased amounts of cortisol, thereby lowering the anxiety, aggression, and stress response.

The most controversial area in epigenetics has to do with later generations being affected by stress and abuse today. When male mice are separated from their mothers after birth, they can suffer from anxiety and features of depression, like listlessness, that are then passed on to subsequent generations. The negative epigenetic changes are actually found in the mice's sperm following the mice's separation from their mothers—the sperm then serving as the vehicle for transmission to the offspring. Related studies have shown that a whole host of effects, from poor diet and stress to exposure to toxins (for example, pesticides that lead to epigenetic modifications in the brains and sperm of mice), can then be transmitted to the next generation.

A profound example of how we may be able to affect our own gene activity comes from a study straight out of science fiction. A Swiss-French team in Zurich was inspired by an innovative game called Mindflex, which comes with a headset that picks up brain waves from the player's forehead and earlobes. By focusing on a light foam ball, the player can lift it up or down on a column of air. The game consists of being able to move the ball through an obstacle course, using thought alone.

The researchers wondered if the same approach could alter gene activity. They devised an electroencephalograph (EEG) helmet that analyzed brain waves and could then transmit them wirelessly via Bluetooth. As reported by *Engineering & Technology* (*E&T*) magazine in November 2014, the brain waves were turned into an electromagnetic field inside a unit that powered an implant inside a cell culture. The implant was fitted with a light-emitting-diode (LED) lamp that emitted infrared light. The light then triggered the production of a specific desired protein in the cells. One of the lead researchers commented, "Controlling genes in this way is completely new and is unique in its simplicity."

The researchers used infrared light because it doesn't harm cells while yet penetrating deep into the tissue. After remote brain transmissions worked on tissue samples, the team progressed to mice, where it was also successful. Various human test subjects were asked to wear the EEG helmet and to control the production of proteins in mice simply by using their thoughts. Out of three groups, the first were made to concentrate their mind by playing Minecraft on a computer. As reported in the *E&T* article: "This group only achieved limited results, as measured by the concentration of the protein in the bloodstream of the mice. The second group, in a state of meditation or complete relaxation, induced a much higher rate of protein expression. The third group, using the method of biofeedback, was able to consciously turn off and on the LED light implanted in the body of a test mouse."

Beyond the amazing implications for the influence of thought directly on gene activities, this approach could someday be applied to help patients with epilepsy by instantaneously delivering drugs or switching certain genes on or off in sufferers via a brain implant at the very onset of a seizure. Just before a seizure, the epileptic brain generates a particular type of electrical activity that could be used to activate a light-activated genetic implant to rapidly produce an anti-seizure drug. A similar strategy could be employed to treat chronic pain by producing painkilling drugs in the brain as soon as the first signs of pain occur.

All in all, our genome is a fantastically nimble assembly of DNA and proteins that is constantly being remodeled in terms of structure and gene activity—and much of this remodeling appears to be in response to how we live our lives. But the Russian-doll problem cannot be swept aside. It's apparent now that chemically induced switches are at the root of shifting gene activity. That much is indisputable. A switch in gene activity in response to one's lifestyle can be brought on by a small methyl group stuck on a gene, leaving a telltale mark. Without this chemical modification of the gene, a stem cell might not develop into a particular brain cell instead of a liver or a heart cell. Indeed, it might not even develop into anything at all but just keep dividing over and over again, the way a cancerous tumor forms.

Methyl marks are not only chemical modifications turning off gene activity but are also like musical notes representing the symphony of more complex gene interactions. By reading the marks as a group, we can get a sense of networks of activity that correspond to how we (and perhaps our parents and grandparents) lived. It might be possible to read directly from the epigenome the specific experiences involved, like living through a famine. Looking at the marks as the score of a symphony makes sense because it takes a multitude of notes before music can really be grasped. Looking at one bar of a symphony provides only a snapshot. Likewise, trying to find the smallest Russian doll doesn't tell you the whole genetic story.

In genetics, the marks are being deciphered chemically, but the step of connecting them to what they mean in terms of experience faces major challenges. First, we can't actually observe genetic changes in real time. Second, we can't connect experience A to genetic change B with any specificity, except in a few cases. It should be possible to find epigenetic alterations from cigarette smoking, for example, yet even then, as we say, not everyone suffers the same damage from the effects of smoking. While we know how chemical marks on certain genes can come about, we cannot say how a certain type of life experience (e.g., prolonged famine) causes specific marks to appear on specific genes in exquisitely precise areas of the genome.

Presently, the biggest challenge remains the missing connection between marks and meaning. When a violinist sees the marks that begin Beethoven's Fifth Symphony—the familiar *ta-ta-ta-DUM*—he goes into action, moving his arm up and down across the strings of the violin. You can see his arm move, but behind this action lie many invisible elements. The violinist knows what the notes stand for, having learned to read music. They aren't just random black-and-white marks on a page. His mind turns the notes into highly coordinated actions between brain, eye, arm, and fingers. Finally, and hardly ever mentioned because it's so obvious, a human being, Ludwig van Beethoven, was inspired to write the symphony and invented the four-note motif known to the whole world. Hundreds of bars of music are based on this simple group of notes.

Even with this knowledge, how does the chemical choreography of millions of genes and their chemically controlled on/off switches deliver the amazing ability of a brain to think? No one knows. How did the brain somehow evolve over eons in response to programming by newly arising mutations? Darwinian genetics would say that all these mutations occurred randomly. But how could this be the whole story, considering that epigenetic modifications in response to how we live our lives may well determine where in the genome new mutations arise? In such cases even Darwin would have to admit that not all mutations occur randomly.

Of course, Darwin could have no idea about epigenetics in his lifetime. But what if he did? Darwin might then tell us that our evolution involves the interplay of both epigenetic marks and new gene mutations. Darwin shocked his contemporaries by excluding God, or any mindful Creator, from his explanation of how modern humans came to be. Certainly, in the study of genetics, assuming some kind of higher intelligence behind the scenes doesn't help us to understand how we evolved. But we can now consider an inherent organizing principle in the evolutionary process that transcends the single-minded concept of random mutations and survival of the fittest. In constructing a new model of evolution, methyl marks on thousands of genes and their histone partners, working hand in hand with the genome, would be helping determine where new mutations will arise (also by influencing the three-dimensional structure of DNA). Then Darwin's natural selection can take over to decide which new mutations persist. In this intriguing albeit speculative scenario, we aren't just blowing in the wind waiting for random mutations to arise. We are directly influencing the future evolution of our genome based on the choices we make.

A NEW POWER PLAYER: THE MICROBIOME

Genetics is in the midst of a knowledge explosion. The data pouring in about the genome and epigenome are piling up every day, not in gigabytes but in terabytes, meaning a trillion bytes of digital information, a thousand times larger than a gigabyte. This mountain of data is difficult to fathom, much less analyze, and another Himalayan range was added from a direction no one anticipated: microbes. In medical school, microbes are chiefly seen as invaders, the bacteria and viruses that cause disease when they break through the body's immune defenses. Off to the side, as it were, friendly microbes were also pointed out, those that live in the intestinal tract, serving to digest the food we consume.

A physician who specializes in gastrointestinal medicine becomes very familiar with what can go wrong in the gut, but most people have very little awareness of the microbes that live side by side with our own cells. Antibiotics, whose purpose is to kill disease-causing germs, also attack the friendly flora in the gut. Normally, these friendly flora restore themselves after a short period once the antibiotic is gone, and the most you'd notice is a bout of diarrhea. When travelers suffer intestinal upsets like "Delhi belly" in India or "Montezuma's revenge" in Mexico, the cause is a change in the ecology of the intestine. Digestive microbes are different in different parts of the world. Unless you feel pain, discomfort, bloating, diarrhea, or constipation, you're not likely to pay much attention to your digestion, certainly not at the microbial level.

In the past few years, however, the whole population of microbes that inhabit us has assumed enormous importance, almost out of the blue. We hinted at the reason in passing when we mentioned that the body contains 100 trillion foreign or microbial cells. As we've already mentioned, this means that 90 percent of the body's cells are microbes, including a vast preponderance of its genetic material. Your body contains about 23,000 human genes, in contrast to over 1 million bacterial genes. In a word, we are a collection of bacterial colonies with a few human cells hanging on! This realization dawned when it became possible to map entire genomes, including the genomes of the hundreds and thousands of possible microbial species that inhabit the body, chiefly in the gut but also on the skin, in the mouth, and in other locations.

Before we can understand our own genes, it's necessary to grasp the genetic implications of the *microbiome*, the label given to the total ecology of micro-organisms, which outnumber our cells 10 to 1 (*microbiota* is also used as a synonym). These microbes didn't just drop in for a visit when higher life-forms appeared. The symbiotic relationship between the cells of our body and trillions of microbes spans vast periods of time, beginning with the first appearance of microbes 3.5 billion years ago; the emergence of our hominid ancestors around 2.5 million years ago represents the blink of an eye in the evolutionary march of bacteria, which can create genes and even exchange them. Along the way, our interaction with these bacteria influenced the evolution of every organ, including the brain. It hasn't been determined how many species of microbes are present in our body; the general estimates rise to more than a thousand—in any case, a bewildering multitude. The impact of the microbiome is suggested by the ways it's been described: "the second human genome";

"a newly discovered organ"; "a bacterial inner rain forest." In the gut, cells are shed in vast numbers: around 100 million to 300 million are shed by the colon *per hour*, a small fraction of the 1 to 3 billion shed by the small intestine. Microbes establish themselves in the biofilm coating the intestinal wall, but they are also shed in quantity—a stool sample contains about 40 percent microbes by weight.

The term *microbiome* was introduced by a Nobel Prize–winning molecular biologist and past colleague of Rudy, Joshua Lederberg, but the notion of a microbiome was first described by a U.S. Army surgeon in the nineteenth century, William Beaumont (1785–1853), a pioneer in the physiology of digestion. He proclaimed that anger hinders digestion. We have since learned that the vast array of gut bacteria directly affects the development of the brain and the central nervous system from the womb to death. In addition, your microbiome is adjusting your immune system every day.

When the natural balance of the microbiome becomes disrupted and unbalanced, we call it *dysbiosis*, yet only now is it being discovered that far from being just a digestive problem, dysbiosis is systemic in the damage it causes. The range of disorders linked to it is growing but is already startling in its numbers: links have been found to asthma, eczema, Crohn's disease, multiple sclerosis, autism, Alzheimer's disease, rheumatoid arthritis, lupus, obesity, cardiovascular disease, atherosclerosis, cancer, and malnutrition. Avenues for new treatments are leading down the same road—to the microbiome.

Excitement over the microbiome has reached such a fever pitch that you've probably become aware of it through the media and through products called probiotics (the most widely advertised is active yogurt), which are advantageous in promoting the growth of healthy microbes in the intestinal tract. From a genetics viewpoint, the microbiome helps to educate the immune system and to prevent disease. Over the eons of evolution, microbial DNA hasn't simply lived side by side with the DNA inside living creatures but has infiltrated it, becoming an integral part of human DNA today. A world of possible findings stems from this interdependence, which has been continuing for millions of years in our species.

The other important story, the connection between the microbiome and chronic diseases, is likely to have a major impact on everyone's life, and it could come quite quickly. There is a natural connection with disorders of the intestinal tract such as irritable bowel syndrome. Obesity is also a natural fit through how food gets digested and metabolized. Far more unexpected is the potential link between the microbiome and far-flung disorders like heart disease, type 1 diabetes, cancer, and even mental illnesses like schizophrenia.

It's now known that gut bacteria produce neuroactive compounds that interact with brain cells and which can even control the expression of our own genes through epigenetics. Once it was realized that there is a strong gut-brain connection, the barriers between our own cells and foreign cells began to crumble. If a bacterium in your intestine can actually influence your mood or contribute to mental illness, a totally new conception of the body looms on the horizon, as we'll explain. (We will discuss probiotics and other dietary recommendations in Part Two, "Lifestyle Choices.")

FROM MYSTERY TO CRAZE

Because hundreds of microbes inhabit your body, their genomes and the terabytes of data derived from them pose a huge mystery. To help make sense of it, we need some general categories to wrap our minds around. Professor Rob Knight, an expert in human microbes at the University of Colorado, introduces the microbiome by saying, "The three pounds of microbes that you carry around with you might be more important than every single gene you carry around in your genome." By weight, the microbiome roughly equals the brain. Knight simplifies the teeming population of micro-organisms by clustering them in the primary areas they occupy around the body, the chief ones being the intestines, skin, mouth, and vagina. These are like separate microbial—and genetic—landscapes, as distinct in their ecology as the Arctic is distinct from the tropics. Behind this simplified map lies Knight's analysis of the microbiome of 250 healthy adult volunteers, and behind that the huge database of genome sequencing by the $173 million Human Microbiome Project funded by the federal government.

One of the chief mysteries about the microbiome is that it varies so much from person to person. In a February 2014 TED talk that has accumulated over 300,000 viewings, Knight tantalizes with some intriguing facts. Some people swear that they get bitten by mosquitoes much more than other people, while some claim they are rarely bitten. The reason partly has to do with the different microbes on their skin and how much mosquitoes are attracted to them. Microbes in the intestine also seem to determine whether an over-the-counter pain medication like Tylenol (acetaminophen) might cause liver damage.

Diversity makes it difficult to describe the population of a perfectly healthy microbiome. On the negative side, modern guts may be severely compromised. In an influential 2014 paper, Stanford University microbiologists Erica and Justin Sonnenburg sounded a message about the possible loss of gut microbes due to various factors. One is the Western diet low in vegetable fiber. Fiber is a *prebiotic*, a food that microbes need to feed on if they are to flourish (as opposed to a *probiotic*, which introduces new microbes into the digestive tract). The widespread use of antibiotics also has a destructive effect on a spectrum of bacteria and viruses. Less tangible but suspicious is our modern stressful lifestyle, because stress hormones and emotions in general can cause shifts in the microbiome. Like your gene activity, your microbiome is so dynamic that it should be thought of as a verb, not a noun.

The most disturbing suggestion by the Sonnenburgs is that modern Western diets are crucial in the rise of chronic diseases and especially autoimmune disorders like allergies. The microbiome helps regulate immunity, and it also produces chemical by-products during the digestive process that reduce inflammation. More and more evidence is mounting that links inflammation to a host of disorders, including heart disease, hypertension, and various cancers. Reducing the diversity of the gut's ecology could be steadily ruining our health. The Sonnenburgs explicitly declare the risks: "It is possible that the Western microbiota is actually dysbiotic [harmful to microbes] and predisposes individuals to a variety of diseases."

As with many issues surrounding the microbiome, these risks are very difficult to validate with total certainty. There are only a few isolated populations around the world whose microbiome is free of damaging influences. Emily Eakin, writing for *The New Yorker* in December 2014, cites the Hadza tribe in Africa, who have been studied by Jeff Leach, an anthropologist who is collaborating with the Sonnenburgs. Three hundred Hadza, who still live in hunter-gatherer conditions in Tanzania, were Leach's subjects for a year. "We need to go to places where people don't have ready access to antibiotics, where people still drink water from the same sources that zebra, giraffes, and elephants drink from, and who still live outside," Leach told Eakins. These are the conditions in which the genes of *Homo sapiens* developed.

Using stool samples, Leach found that "it looks like the Hadza have one of the most diverse gut ecosystems in the world of any population that's been studied." Yet a previous study of the Hadza by researchers from the Max Planck Institute for Evolutionary Anthropology in Germany disclosed that while they harbored certain gut bacteria never seen before, the Hadza lacked others that are associated with good health in the Western microbiome. Leach believed enough in the genetic superiority of the Hadza gut, however, that he transplanted a sample of their microbiome into his intestinal tract.

This leads to a craze that has gone viral despite the fact that so much of the microbiome is up in the air. The way that Leach transplanted the Hadza's microbes was by using a turkey baster to inject their feces into his colon. However distasteful, even repellent, this sounds, there are YouTube videos instructing how to do the same to yourself. The basis for this DIY procedure is simple logic. If a Western adult microbiome is compromised, that of a newborn baby or healthy young child isn't. Why not exchange one for the other?

The U.S. Food and Drug Administration (FDA) has stepped in to prevent fecal microbiota transplantation (FMT) by doctors until official trials are conducted along the same lines as introducing a new drug. Such is the enthusiasm for FMTs, however, that the practice has gone underground, and there are no prohibitions on its use by doctors in other countries. The FDA's ruling put an immediate stop to small-scale research that lacks the financing to pursue exorbitantly expensive trials lasting typically from seven to ten years. But the FDA, as Eakins reports, was stung by the thousands of people who developed acquired immunodeficiency syndrome (AIDS) through blood transfusions before it was known that blood was a means of transmitting the human immunodeficiency virus (HIV). Disease organisms like the virus that causes hepatitis A are harbored in the gut. (In the case of hepatitis A, infected fecal matter must enter the mouth of someone who isn't immune to the disease; this usually happens in unsanitary conditions with food handlers.) This and other risks as yet unknown make the FDA ruling appropriately cautious.

Performing an FMT is like taking the donor's entire microbiome without knowing what it contains. No one should take such a risk. But the underground fad for FMT, messy and off-putting as the whole procedure is, rests on the enormous potential of the microbiome to reverse so many chronic illnesses. One striking example is Crohn's disease, an inflammatory bowel disease that can be totally debilitating. Symptoms include chronic diarrhea, which can lead to severe weight loss, along with abdominal pain and fever. Victims of Crohn's disease tend to lead miserable lives, held captive by their illness. Because the root cause is inflammation of an unexplained origin, there can be inflammatory problems outside the intestinal tract as well, such as skin rashes, red and swollen eyes, and even diabetes.

Drug therapies are often ineffective in Crohn's disease, and in severe cases the most damaged sections of intestine are surgically removed. But going back to the 1950s, isolated physicians, generally considered renegades or worse, believed that treating Crohn's patients with fecal matter from healthy donors (taken under sanitary conditions by pill or through the rectum) produced actual cures, often in a remarkably short period of time, meaning weeks and months. Now treating Crohn's disease through FMT could go mainstream, and even the FDA made an exception for it in their ruling against the procedure.

Even more startling is a condition that FMT seems to cure in a matter of hours, even when a patient is close to dying. The condition is a bacterial infection of *Clostridium difficile*, which arises in connection with strong doses of antibiotics. Up to half a million people currently suffer from the infection, with more than ten thousand dying annually in severe cases. *C. difficile* resists antibiotics and typically is found when a hospitalized patient being treated with a heavy course of antibiotics has had severe depletion of their microbiome. The conditions are then ripe for *C. difficile*, with the infection causing symptoms similar to Crohn's disease, including severe diarrhea.

Ironically, the standard treatment for *C. difficile* is to give vancomycin, an antibiotic. Vancomycin can be totally ineffective if a new, resistant strain of the bacteria has emerged. But the medical literature contains scattered reports of remarkable, almost instant recovery using FMT. Within hours, the newly inserted microbes defeat and crowd out *C. difficile*, leading to a subsiding of all symptoms. The FDA has also made an exception in this case. By extension, if an FMT can heal two disorders that share the same symptom—highly destructive inflammation—and if inflammation is potentially the villain in chronic diseases of many types, why not take a chance and perform our own FMT, using the healthiest stool you can convince someone to donate? This is the logic that made home FMT go viral.

No one has proved that taking such a step is either good science or effective medicine, and we are certainly not condoning it. (There are other, safer ways to optimize your microbiome, as we'll see.) But findings in animal trials indicate that a true revolution may be brewing. In 2006 a team from Washington University in St. Louis apparently proved a strong connection between the microbiome and obesity. They took mice that had been genetically altered to be obese and transferred some of their microbes to normal mice. Those mice became obese, which is the first time that a disorder has been transferred via the microbiome, at least in animals. But what's truly startling is that the mice who grew fat after receiving the microbes ate the same diet as mice without the transplant, and yet the untreated mice didn't grow fat.

How did the same caloric intake produce fat mice and normal mice at the same time? It's presumed that the inserted microbes were somehow more efficient at extracting nutrients from the food as it was being digested. This runs counter to a long-held belief that calories in equal calories out. In other words, if a meal contains a thousand calories, everyone's body, given complete, healthy digestion, will extract a thousand calories of energy. Yet we all know people who say, "I only have to look at a piece of chocolate cake to gain a pound." Provocatively, this new study suggests that they have a point. Some microbiomes may work better at nutrient extraction than others, with obese people extracting too much and skinny people extracting too little.

Researchers in Amsterdam wanted to see if a fecal transplant of microbes from lean people into fat people would be enough to cause them to lose weight. So far, it hasn't been. The subjects showed improved insulin sensitivity (key to whether calories are properly metabolized instead of being stored as fat), but they didn't lose weight, and the benefit was gone after a year. It may be that more treatments are needed or that specific microbes need to be isolated from "lean" microbiomes. The whole genetic story has yet to be told, however, and it could prove far more complicated.

ENTERING A NEW ECOLOGY

As you can see, adjectives like *foreign*, *alien*, and *invasive* are not applicable to the microbes that have learned to cooperate with the human body over millions of years. There are indications that a baby's normal development may depend on them. Going back to Professor Rob Knight's simplified map, there are different microecologies located in the mouth, intestines (feces), skin, and vagina. Before a baby is born, its body has no microbes; the gastrointestinal (GI) tract is actually sterile. As it passes through the vaginal canal, the baby receives a filmy coating of the mother's microbiome from that area. Birth is just the first step in exposing a baby to microbes, which it will take in from every direction: the mother's breast, food, water, air, pets, and other people. The GI tract begins to be colonized within hours after birth. Animal studies have shown that when raised in a sanitized, microbe-free environment, animals develop a range of abnormalities from immune deficiency and shrunken heart to improper switching in brain cells, along with the expected digestive problems.

Sometime in childhood the microbiome ceases to be in constant flux. It stabilizes, although not in the same way for everyone. On Knight's chart, the progress of the early microbiome moves from the skin-vagina region at birth to the intestinal-fecal region. This sequence is true for everyone, because a gut that can digest food is universal. But there is evidence that the more exposure to microbes, the better, which is a kind of paradox. Children raised in developing countries exhibit much more diversity in their microbiome, furthering the probability that we live "too clean" in the developed West. But such children also suffer from more childhood diseases, just as kids who are left in day care may, it seems, be less prone to allergies but also run the risk of catching more colds, earaches, flu, and other transmissible diseases.

In epigenetics, as we've seen, the biggest problem is the absence of straight-line cause and effect. A doesn't lead to B when a cloud of causes presses in on the mind-body system. With the microbiome, the big problem is how rapidly it changes. Genes are far more stuck, even taking the epigenome into account, than the microbes that inhabit us. Imagine the ocean shoreline as waves beat against the land, constantly moving the sand. Tides and weather determine how much sand is taken away or deposited. If the grains of sand were living microbes, the tides and weather of the gut are constantly moving microbes around, flushing some out and allowing others to enter.

Using the word *ecology* may sound like a metaphor, but medicine is only beginning to understand that the intestinal tract, which is roughly twenty-five feet long and has a surface area comparable to a tennis court, is as complex and dynamic as the global ecology. It's estimated that the microbiome has somewhere between 40 and 150 times more genes than the body itself. As an example of what surprises lie in store for explorers of this ecology, let's consider a disorder that is now being strongly connected to microbes: obesity.

The timeworn model of "calories in, calories out" puts the onus for obesity on a person's eating habits. If you eat too much, for whatever reason, your body stores the excess calories as fat. Studies do in fact show that overeaters tend to underestimate how many calories they consume. But if overeating were the only cause of obesity, it doesn't explain why only 2 percent of dieters successfully take off at least five pounds and keep it off for two years. What's forcing their hand? One possibility is the stuckness of bad habits, which causes the old eating patterns to creep back into a dieter's life. But weight gain has been associated with a variety of influences. The following list isn't meant to alarm or depress you, only to illustrate how complex the natural activity of eating has become.

Why People Gain Weight

They overeat.

They come from a family of overeaters, with a possible genetic connection.

Their friends overeat.

Their diets contain too much refined sugar, simple carbohydrates, and fat.

They consume too few fresh fruits and vegetables and other sources of soluble fiber.

They eat processed, junk, and fast food that contains additives and artificial ingredients, along with excessive salt and sugar.

They develop a range of bad eating habits: watching TV while eating, eating too fast, snacking between meals, and so forth.

Their lives are stressful.

They are undergoing a personal crisis, such as being fired or getting a divorce.

There is an imbalance between the two hormones (leptin and ghrelin) responsible for making someone feel hungry and full.

Their brains show inflammation or damage to the hypothalamus, the center for regulating appetite.

Their bodies exhibit signs of chronic inflammation.

They've given up on losing weight after years of yo-yo dieting.

They recently quit smoking and overeat to compensate.

With so many factors at work, usually in concert, it becomes abundantly clear why obesity remains difficult to treat. One disorder overlaps the separate fields of nutrition, endocrinology, genetics, gastroenterology, psychiatry, and sociology, each having its own perspective. The cloud of causes looms heavily. Yet through all these complex influences one thread may be pulled out: the microbiome, which primarily digests food but also exerts a major effect on hormones, immunity, stress response, and chronic inflammation. There is no other factor that encompasses so many bodily functions.

The trail of clues leads from food to gut to the whole body. Someone who has followed the trail is Dr. Paresh Dandona, a diabetes specialist at the State University of New York at Buffalo's School of Medicine. Dandona had a major clue fall into his lap when curiosity led him to examine the food at McDonald's. Nine volunteers who were of normal weight consumed typical McDonald's breakfast fare: an egg sandwich with cheese and ham, a sausage muffin sandwich, and two hash brown patties, which totaled 910 calories. There are well-known reasons besides calories why such a breakfast, high in fat and salt while containing almost no fiber, is unhealthy. Dandona added something quite unexpected. As reported in *Mother Jones* magazine in April 2013:

> Levels of C-reactive protein, an indicator of systemic inflammation, shot up "within literally minutes. . . . I was shocked," [Dandona] recalls, that "a simple McDonald's meal that seems harmless enough"—the sort of high-fat, high-carbohydrate meal that 1 in 4 Americans eats regularly—would have such a dramatic effect. And it lasted for [five] hours.

Using a phrase like "harmless enough" reflects the somewhat laissez-faire attitude taken toward fast food by many Americans. (Besides causing an upsurge in inflammation, consuming a Big Mac rapidly injects fats into the bloodstream that can be observed as visible clouding in the serum [clear liquid] after the red corpuscles are centrifuged out.) Dandona's research took a major turn, and he made even more startling discoveries.

Over the next decade Dandona examined various foods to see how they affected the immune system, which is known to be compromised by chronic low-level inflammation. Reporter Moises Velasquez-Manoff writes, "A fast-food breakfast inflamed, [Dandona] found, but a high-fiber breakfast with lots of fruit did not. A breakthrough came in 2007 when he discovered that while sugar water, a stand-in for soda, caused inflammation, orange juice—even though it contains plenty of sugar—didn't." Somehow fresh, unprocessed orange juice counteracted even the 910-calorie McDonald's breakfast splurge. Among the test subjects, the breakfast caused inflammation and elevated blood sugar whether they accompanied the meal with sugar water or plain water. Yet neither effect showed up among the subjects drinking orange juice.

Velasquez-Manoff continues. "Orange juice is rich in antioxidants like vitamin C, beneficial flavonoids, and small amounts of fiber, all of which may be directly anti-inflammatory. But what caught Dandona's attention was another substance." This was a molecule called endotoxin (literally "inner poison") that appeared in blood after eating the McDonald's breakfast among the subjects drinking water and sugar water but not among the orange juice group. Endotoxin is produced by the outer membrane of bacteria, and its presence in the bloodstream signals the immune system to go into action, with resulting inflammation. Dandona suspected that the source of the endotoxin was the microbiome. The endotoxin got into the bloodstream by being ushered there through the intestinal wall by the McDonald's food. The orange juice somehow kept the endotoxin inside the gut, where it's naturally found. (More research on "leaky gut syndrome" is deepening the connection to diet.)

Orange juice isn't a panacea or unique in its effect; there could be a wide range of foods that counteract chronic inflammation. In the face of an ever-shifting microbial ecology, some constant influences may be enough to change the course of a person's well-being. But what's needed is more than a kitchen cupboard of beneficial foods, important as that is. (See pages 121–122, where we recommend the best microbiome diet, as far as current research indicates.)

FROM CLUES TO CASCADES

Dandona's findings, among others, do more than reinforce the standard recommendation that a balanced diet should contain soluble fiber from whole fruits and vegetables, along with whole grains. The prospect of reversing unhealthy inflammation is exciting. Advances come from unexpected places. It's been observed that the inflammatory molecule endotoxin decreases in the bloodstream after someone undergoes gastric bypass surgery. Gastric bypass is a procedure that reduces the stomach to a small pouch about the size of an egg. The small intestine is directly connected to this pouch, and as the result of having a severely reduced stomach, patients eat less and therefore can lose dramatic amounts of weight.

That was the accepted explanation, except that the reduction of inflammation points to the microbiome. In a series of tests with rats and mice, a team from Massachusetts General Hospital produced a remarkable result. They performed gastric bypass on the rodents, and afterward their microbiome completely reset itself. A surge of beneficial microbes not only reduced inflammation but led directly to weight loss. This cause-and-effect sequence was shown by taking the microbes from these gastric-bypass animals and inserting them into the intestines of germ-free mice. The injected mice lost weight while still eating their previous high-caloric diet. In fact, they lost weight while consuming more calories than did a control group of mice that lost no weight. This result helps to debunk the long-accepted belief that weight gain and weight loss are entirely about calories. It also points to another intriguing possibility. As part of the microbial resetting, the gastric-bypass and injected mice were able to metabolize glucose, or blood sugar, in a normal, healthy fashion, which wasn't true of the mice that lost weight by eating less. Considering how human dieters almost always regain the weight they lose, it may be that the problem isn't about returning to the "wrong" diet, losing willpower, or secretly consuming too many calories. It may be, as with these mice, that it takes a reset of metabolic processes controlled by the microbiome.

We'll go into this subject in depth in Part Two, which covers lifestyle changes, but it's worthwhile to summarize the possibilities here.

What Would Reset Your Microbiome?

Eating less fat, sugar, and refined carbohydrates
Adding sufficient prebiotics on which bacteria feed: fiber from whole fruits, vegetables, and grains
Avoiding chemically processed foods
Eliminating alcohol consumption
Taking a probiotic supplement (see page 124)
Eating probiotic foods like yogurt, sauerkraut, and pickles
Reducing foods with inflammatory effects
Focusing on foods with anti-inflammatory effects, like freshly squeezed orange juice
Diligent stress management
Attending to "inflamed" emotions like anger and hostility

We want to emphasize that these are all *possibilities* rather than certainties. The microbiome reaches beyond digestion into every part of the body. Therefore its effects are extremely complicated, and further research is continually needed. What's known so far, however, looks very promising.

For example, much disease seems to be the result of a cascade of processes in the body, meaning a chain of events that follow one upon the other, creating more problems as the cascade progresses. For example, mice raised without their normal complement of microbes can gorge on food without gaining weight, owing to inadequate digestion. But if put back in with other mice so that they acquire a normal microbe colony, the gorging mice run into trouble. The excess calories are now being digested and have to be stored as fat. Their livers become insulin resistant, and the animals become obese even on fewer calories.

The same cascade can also be produced through endotoxin. Belgian researchers led by Professor Patrice Cani gave mice small doses of endotoxin, which caused their livers to become insulin resistant. Obesity followed and then diabetes. This sequence pointed to the possibility that leaks from the microbiome might be a major factor in human obesity, exacerbated by overeating and eating the wrong foods. "Then came the bombshell," Velasquez-Manoff writes. "The mere addition of soluble plant fibers called oligosaccharides, found in things like bananas, garlic, and asparagus, prevented the entire cascade—no endotoxin, no inflammation, and no diabetes." Cani had found a means of preventing the damage by something equivalent to Dandona's orange juice: fiber. When certain soluble fibers are intact when they reach the colon, where the bulk of digestive microbes live, the bacteria break the fiber down as food. Thus a prebiotic, a necessary precursor to a healthy microbiome, stopped the disease cascade in its tracks. Fiber is noncaloric, but as microbes break it down, beneficial substances are released, including acetic acid, butyric acid, B vitamins, and vitamin K. (Also worth recalling are the mouse trials at Washington University, in which transplanting microbes from obese mice caused normal mice to become obese without overeating.) Following is a summary list of the implications of this research into the gut-inflammation connection.

The Gut-Inflammation Connection

Fatty, high-carbohydrate foods promote inflammatory substances in the bloodstream.

Endotoxin and other harmful molecules released by certain bacteria can leak through the intestinal wall.

If such leakage takes place, an immune response is triggered, and inflammation results.

Inflammation disturbs, among other things, blood sugar levels and the liver's insulin response.
When that happens, obesity may occur even on a diet containing a normal amount of calories.
Orange juice and soluble fiber shift the balance toward a beneficial microbiome and counteract the cascade that follows from a "leaky gut."

Many researchers now feel that the gut-inflammation connection has uncovered a major source of chronic disease, not just obesity. Links to diabetes, hypertension, heart disease, and cancer are being vigorously pursued. "If we take care of our gut microbiota, it will take care of our health," Cani says. "I like to finish my talks with one sentence: 'In gut we trust.' "

When you explore the fast-mounting studies on the microbiome, the gut-inflammation connection becomes even more important. Liping Zhao, a Chinese microbiologist, told his own story to the journal *Science* in June 2012, part of a special issue devoted to the microbiome. In "My Microbiome and Me," Zhao presented himself as a human guinea pig who reversed his own obesity, high level of "bad" cholesterol, and elevated blood pressure by switching to a diet strong in whole grains along with two foods taken to be beneficial in Chinese medicine, bitter melon and Chinese yam. Losing forty-four pounds in two years is impressive, but Zhao had suspected a connection between obesity and inflammation in 2004. It seems highly significant that in his own case, one microbe, *Faecalibacterium prausnitzii*—a bacterium that has anti-inflammatory properties—flourished in his gut, increasing from an undetectable percentage to 14.5 percent of Zhao's total gut bacteria.

The changes persuaded him to focus on the microbiome's role in his transformation. Mouse trials followed, then human trials. One patient who was morbidly obese, weighing 385 pounds at age twenty-six, experienced many of the same benefits as Zhao, shedding over 100 pounds in a year. Once more there was a specific microbe involved. A single bacterium, *Enterobacter cloacae*, known to create inflammation, made up more than one-third of the patient's microbiome. In this patient, on Zhao's diet it dwindled to trace amounts, while anti-inflammatory microbes increased.

Targeting specific disease processes and "bad" microbes may not be necessary in reversing obesity. One study looked at four pairs of identical twins in which one twin was lean and the other fat. Mice received gut microbes from one or the other twin, and the mice that received the microbes from the fat twin became obese, with a thicker layer of fat. We'll look at the implications of this key finding for your own diet in Part Two, on lifestyle.

This book is about genes, not the microbiome, but it's now impossible to discuss genes without it. Your microbiome is, in essence, your second genome. But unlike your own genome, your microbiome is contagious, because you can spread your bacteria to others. And while this may sound a bit gross, the exchange of bacteria between people through intimate contact can benefit the population. Some evolutionists have gone so far as to propose that human social behavior basically evolved to promote the sharing of microbes. Increasing resistance to infections and dietary toxins could be a dominant factor. In vegetarian species of animals, the microbiome is primarily for digesting a plant diet, but raw meat from a lion kill, for example, is likely to be filled with parasites, disease organisms, and toxins, so a carnivore's microbiome protects the creature from these. Human evolution picked up from there to maximize our disease resistance to its present level.

THE GUT-BRAIN AXIS

With a wealth of gut genomes so vastly outnumbering our own, the microbiome exerts a powerful influence beyond digestion and metabolism. Most fascinating, perhaps, is the "gut brain." Christine

Tara Peterson, Ph.D., who has examined this area in depth (she is also associated with the Chopra Center, doing advanced research on the microbiome), points out that the gut harbors 100 million neurons, more than the spinal cord, and produces 95 percent of the body's serotonin, one of the most crucial neurotransmitters, whose levels are connected, it's long been thought, to depression.

The brain's main line of communication to every region of the body is via twelve cranial nerves. One is the vagus nerve, named for the Latin word for "wandering." Its wanderings are extensive, beginning at the medulla oblongata in the lower brain, passing down the neck, past the heart, and into the digestive tract. Around 80 percent of all the sensory information that reaches the brain is transmitted via the vagus nerve as it branches out. What's intriguing for our purposes is that 90 percent of the traffic, Peterson says, is from gut to brain. "The microbiome," she points out, "may be impacting mental states like anxiety or autism."

The clues are hard to follow, however, because few labs are set up to follow the trail of molecular messages from the gut to the brain. But it's accepted that gut-brain access is a two-way street. The bacteria in your intestinal tract affect the workings of your brain, having a potential to alter emotions, even the risk for neurological and psychiatric disease. In turn, your mood and stress level affect the bacteria that will live in your microbiome. What's come to fruition is an idea proposed early on by the eminent psychologist William James, working with a physiologist, Carl Lange, in the 1880s. They held that emotions arise because the brain is interpreting signals or reactions from the body. In updated form, this has turned into a feedback loop between brain and body using chemical messages.

Beginning as far back as 1974, studies of baby monkeys have shown that separation from their mothers at birth is more than psychologically distressing—it changes their gut microflora. In a related study in which baby mice were separated from their mothers, they became more anxious compared with those that stayed with their mothers. Yet when the intestinal tracts of the maternally separated mice were recolonized with bacteria from the mice that stayed with their mothers, the anxiety of the separated mice dissipated. These results apparently extend to humans as well. If the gut bacteria from human patients with irritable bowel syndrome are placed in the gut of mice, the mice become socially inept and anxious. Emotional distress has long been associated with irritable bowel syndrome, and now it seems that the connection has a material basis, not simply a psychological one.

In another study, a Dutch team has shown that if new mothers are stressed, their stress actually changes the microbiome of their infants. It seems highly plausible, then, that chronic social stress could be changing your intestinal bacteria, creating a destructive feedback loop between gut and brain that causes inflammation throughout the system, including the brain. It's fair to say that while modern medicine has focused for more a century on killing bacteria, we're now learning to live healthier lives *with* them.

Whether or not you find all this talk about gut bacteria distasteful, anyone can be forgiven for feeling humbled by it. We humans are used to seeing ourselves as above other creatures, certainly above micro-organisms that are the most primitive life-forms on Earth. These microbes have moved from parasites to partners. Theoretical biologist Stuart Kauffman has rightly said, "All evolution is co-evolution," while the pioneering quantum physicist Erwin Schrödinger once declared, "No self is of itself alone . . . the 'I' is chained to ancestry by many factors."

But finding our evolution tied to microbes can be reframed so that it's not at all humbling. Inside our bodies, through our own genome and the genomes of microbes, is contained the entire history of life on Earth. Every person is a biological encyclopedia; every generation writes a new page or chapter. Since the body you see in the mirror is *life itself*, the need to preserve the ecology becomes much more necessary, because the ecology is no longer "out there."

What you eat for lunch today is on the same level as saving the rain forest or reducing greenhouse gases, a form of self-preservation that cannot be put off as someone else's problem. In that light, Part Two will describe how a radical redefinition of the body leads to a new lifestyle and the fruit of that lifestyle, radical well-being.

Part Two

LIFESTYLE CHOICES FOR RADICAL WELL-BEING

What makes the new genetics astonishing is that it has caused us to realize something that's easy to forget. Nothing is more remarkable than the human body. It changes dynamically with every experience, responding with perfect precision to life's challenges—if only we let it. Beyond normal health and vitality, your body is the platform for radical well-being. Every cell is prepared for this transformation, powered by the super genome, but our mind hasn't been. Now you have the knowledge in hand, and we hope you've accepted a much more expanded view of possibilities.

You need to awaken these possibilities. As long as people's lifestyles had no genetic consequences, the only proven approach to greater well-being was standard prevention. Now, with two major breakthroughs—epigenetics and the microbiome—our genes can say yes to a broad range of positive changes. Any gene has the potential to become a super gene when it cooperates with our intentions and desires. Personal evolution needs this cooperation, or we can't move forward.

All well-being, whether radical or not, contains two simple steps.

First, find out what's good for you and what's bad.

Second, do what's good for you while avoiding what's bad.

When it comes to the first step, a lack of knowledge—along with a host of mistaken beliefs masked as knowledge—had to be overcome in the new genetics. If you know, as we now do, that only 5 percent or less of disease-related gene mutations are fully penetrant (deterministic), that leaves 95 percent open to change in their activities.

The second step is about implementing your knowledge, and here is where the biggest challenges lie. Standard prevention, with its well-known risk factors and familiar advice, has broadcast the same healthy message for more than forty years. Why, then, aren't people healthier than ever? Cancer death rates have decreased only marginally since the 1930s, despite some dramatic successes with early detection. Smoking remains a problem for 25 percent of the population, and obesity rates keep rising. The devil, it turns out, isn't in the details; it's in the denial.

Deepak attended a conference recently on the benefits of meditation in which the news was remarkably promising. The speaker, a world-famous genetics researcher, was focusing on how meditation produced beneficial gene activity through the epigenome (we'll talk more about the relationship between meditation and your genome later). When the period for questions came, someone in the audience asked, "Given all of these fantastic findings, do you meditate?"

"No," the researcher replied.

The questioner was shocked. "Why not?"

"Because," the speaker said, "I'm looking to develop a pill that will bring the same results."

He got a laugh, but being humorous about your noncompliance leads to the same outcome as other kinds of denial. Motivating people to do what's good for them and to avoid what's bad must be the first order of business. We all contend with the voice in our head that says

I'll get around to it later.
It's too much trouble.

I'm probably all right anyway.
Would it really make that much difference?

The "it" can be anything you know needs improvement—a better diet, regular exercise, stress reduction, and so on. Sometimes denial doesn't need any voice making excuses. A kind of convenient amnesia sets in when we're tempted by a piece of chocolate cake, which we're not even hungry for, or by a favorite TV show that makes us forget to take a walk after dinner.

Let's do a quick spot check on your present situation. Following is a quiz in two parts—the first part is about doing what's good for your genome, the second about avoiding what's bad. We want you to self-assess as honestly as you can. Your answers will serve as a good preparation for the lifestyle choices outlined in this section of the book.

We begin with the lifestyle habits that send positive messages to your genome.

QUIZ (PART 1): THE LIFE YOUR GENES WANT

Put a check beside each item that is almost always (90 percent of the time) true about you.

___ I allow my life to unfold naturally, without a hectic schedule and constant demands.

___ I get sufficient sleep every night (at least 8 hours) and wake up feeling refreshed.

___ I follow a regular but not rigid daily routine.

___ I pay attention to staying in balance with my diet, eating from all the healthy food groups.

___ I avoid toxic food, air, and water, including food loaded with artificial ingredients.

___ I don't skip meals.

___ I don't snack.

___ I take steps to minimize my stress and manage the stresses that are unavoidable.

___ I give myself some time out every day to let my body reset itself.

___ I meditate.

___ I do yoga.

___ I eat moderately and maintain a healthy weight.

___ I avoid long periods of sitting, moving my body at least once an hour.

___ I don't smoke.

___ I drink alcohol sparingly or not at all.

___ I avoid red meat, and if I do eat it, I do so sparingly.

___ I do my best to eat only organic foods.

___ I am physically active.

___ I understand the danger of chronic inflammation and take steps to avoid it.

___ I place a high value on my own well-being and practice self-care every day.

Score: ___________ (0 to 20)

Now assess the negative side, the lifestyle habits that send the wrong messages to your genome.

QUIZ (PART 2): THE LIFE YOUR GENES DON'T WANT

Put a check beside each item that is fairly often (50 percent of the time) true about you.

___ I approach my day as an endless round of things I have to get done.

___ I feel exhausted by the end of the day.

___ I habitually drink to unwind.

___ I am driven to be a success, even though it has personal costs.

___ I get poor or erratic sleep. I wake up still feeling tired.

___ I go to bed with my mind full of thoughts, often worrisome.

___ I smoke.

___ I allow my body to get pretty far out of balance before I tend to it.

___ I don't bother about food labels and the ingredients on the package.

___ I complain about stress but do little to manage it.

___ I am constantly busy and on the run, leaving no time for me to be quiet and calm.

___ My diet is careless.

___ I snack, particularly late at night.

___ My weight isn't where it should be.

___ I don't pay attention to whether food is organic or not.

___ I prefer red meat over chicken and fish.

___ I sit for long periods of time (2 hours or more) without moving, either at work, on the computer, or watching TV.

___ I am considerably less active than I was ten years ago.

___ I worry about aging but don't follow any anti-aging regimen.

___ I don't think much about caring for myself.

Score: __________ (0 to 20)

Looking at your two scores, here's a rough evaluation.

Part 1: On the positive side, if you checked around 10 items, you are living like the average American. Prevention has made an impression on you, but the results are hit or miss. A score less than 10 implies that you are running considerable risk for problems down the line. A score over 15 is very good news—the super genome is already saying yes to your lifestyle.

Part 2: The scoring here is about sending negative messages to your genome more than half the time. If you score a 10, which is probably close to average for how Americans live today, you probably enjoy good health but run the risk of future problems. Even one bad habit has the potential to modify one or more genes in undesirable ways. A score of less than 10 puts you in good shape for moving forward. A score of 12 or more implies that you should urgently consider how to improve your well-being.

RENÉE'S STORY

We'd love it if everyone got a 20 on the first quiz and a zero on the second one. But being realistic, there's always room for improvement. Even though the lifestyle habits we've listed are well known in standard prevention, what's new is the precise and constant attention that the super genome is paying. Nothing escapes its attention. That's great once you decide to make positive changes, not so great if you remain in the same groove. We can illustrate the situation created by the new genetics through one woman's story.

Renée, now in her early fifties, has been steadfast in doing what's good for her. She eats a diet of whole foods from every group (fruits, vegetables, legumes, grains). She never eats fast or junk food and hasn't touched alcohol in years. Every day in summer she swims; when the weather gets cold she takes a brisk walk after dinner. Renée's marriage is good, and she thoroughly enjoys her work as an alternative therapist. Why, then, does she weigh over 225 pounds, having struggled with her weight since her early teens?

Renée's denial is one of timing. When food is in front of her, she has no impulse control and digs in as if she has no weight problem. It's when the meal is over, in all the hours between meals, that she suffers from the realization that her problem is real and not getting better.

Hank would seem to be in a much better situation. He's sixty-five and has no physical problems other than the extra twenty pounds he associates with middle age. Since he has no aches or pains and rarely gets even a cold, he considers himself fortunate compared with many of his friends with their rash of hip and knee replacements. "I can still eat anything," says Hank, who claims to have no digestive problems, which fits in with his claim that he's never had a headache, backache, or stomachache.

His is a subtler form of denial than Renée's. Hank denies that time will bring future problems. Because he feels good today, he ignores almost all disease-prevention advice. He doesn't exercise and sits for long hours a day at the computer, virtually without moving. He eats a wide range of junk and fast food, with frequent snacking. He has no idea what his blood pressure is, having stayed away from doctors for decades. Is he going to be the exception to all the risks he's running?

On the spectrum of denial, most people fall somewhere between these two extremes. Finding the motivation to do what's good for them is hit or miss. Most days they might be careful about what they eat; a couple of hours a week they may find the time for physical activity; sleep problems, if they exist, are generally sporadic. But from our perspective, this situation, which feels normal to millions of people, denies them the possibility of radical well-being. Let's see how that can change.

LESSONS IN CHOICE MAKING

See yourself sitting in your favorite restaurant, feeling relaxed and content. You've eaten just enough, but the waiter comes around with a familiar temptation: "Leave room for dessert?" You don't give in immediately but ask to see the dessert menu. "Coffee, after-dinner drinks?" he asks.

"Let's see," you say, relenting a little bit more. As you glance over the dessert list, there's a pause, which could last only a few seconds, and then you pivot into action. Nothing is more important than this pivot. It's where you call upon a certain aspect of yourself, the choice-making part. Do you give in to temptation or not? Unless you fall into one extreme of total self-discipline or the other of total lack of impulse control, there's no predicting what you will choose.

Choice making is difficult, even when it comes to small daily decisions, and so instead of getting better at it, approaching it as a skill, we behave haphazardly. Between knowing what's good for you and doing it there's a gap. In this gap is where the skill of choice making is learned. Eating a rich dessert and having chocolate remorse afterward comes too late.

Yet if you could make just one significant change a week, your progress toward radical well-being would be hugely accelerated. After a month you would feel some real benefits; after a year the transformation would be complete. Reduced to a steady string of easy choices, the problem of noncompliance would disappear. You can even allow yourself to be in denial without feeling guilty, just as long as you alter one thing a week, whether it's in your diet, your daily routine, or your physical activity. Just deciding to stand up and move around every hour, which seems like a trivial choice, sends positive messages to the super genome, enough to make a difference in gene activity.

The goal of one positive change a week won't be attainable, however, without a workable strategy. If you try to change by making a resolution, you'll fail. Millions of people make New Year's resolutions, which constitute only one change in the coming year, and yet the vast majority, well over 80 percent according to polls, don't follow through on their resolutions for more than a short time. Making promises to yourself, feeling guilty over your lapses, and feeling lonely and self-pitying are all counterproductive. Someone addicted to alcohol or drugs wakes up every morning with these feelings. Their past is littered with broken promises to themselves.

In the welter of advice that repeats the same thing over and over—"Make good choices"—very little advice is given on *how* to do so. Let's consider three basic principles we must deal with in making choices.

1. *There are easy choices and hard choices.*
   Both kinds present themselves every day, but we usually don't stand back and pay attention to which is which. We carry on as usual, driven by habit, old conditioning, and sheer unconsciousness. The hard choices, then, are those that try to move the psychological machinery in a different direction. On the surface, a choice may seem quite small, but big or small isn't the issue. The issue is how hard the choice is. To someone with a severe phobia of insects, picking up an ant or a dead cockroach constitutes a hard choice, and at times an impossible one. On the other hand, soldiers in battle routinely risk their lives, rushing in under heavy fire to rescue a fallen comrade. The objective facts about a choice—whether you are risking a little or a lot, whether the choice is easy for other people or not, whether it will bring pain or pleasure—is

secondary and sometimes totally beside the point. What's primary is whether the choice feels hard or easy to you.

2. *Bad choices sometimes feel good.*
   There's no mystery here. If you want instant gratification, a shot of joy juice can be had from ice cream at midnight or "eating the whole thing." Guilty pleasures provide a double boost by offering gratification while briefly making the guilt go away. The downside, which isn't news, is that the feel-good result starts being less effective, and after a while the guilt is so great that nothing really feels good anymore.
3. *The gratification from good choices is usually delayed.*
   This has become a classic psychological axiom, thanks to a famous test from the sixties and seventies known as the Stanford Marshmallow Experiment. In one version, young children were sat down with a piece of marshmallow candy in front of them. "You can have the marshmallow now," they were told, "but if you wait ten minutes, you will get two marshmallows." The researcher left the room, and the children were observed through a two-way mirror. Some children immediately ate the marshmallow or ate it after a brief struggle with themselves. Other children waited, even when showing signs of struggle, for the delayed gratification.

   From this simple test, some psychologists believe, you can tell much about what kind of adults these children will grow up to be. The instant gratifiers will become prone to impulsive decisions, regardless of the consequences. They may possibly take more risks or ignore the risks in a given situation. Their ability to plan for the future will be diminished. None of this is so surprising if you remember Aesop's fable about the grasshopper and the ant. The real issue is whether the bad habits of live-for-the-moment grasshoppers can be changed.

Anyone should be able to see how these issues are at work in their own lives. If you look back at the three people's stories given as examples of denial, it hardly matters that Ruth Ann, Saskia, and Renée are very different as individuals. The basic principles of choice making apply to all of us. The question is how to use these basic principles of choice making for our own advantage. Following are what we believe are the most workable answers.

1. There are easy choices and hard choices.

The answer to turning this principle to your advantage is to start your transformation with making small, easy good choices. As these good choices accumulate day by day, you will be sending new messages to your epigenome and microbiome, the two great centers of change in every cell. At the same time, each daily change, however small, is retraining your brain. It starts getting used to a new normal. In contrast, hard choices make you run into a brick wall, because the brain can't face a drastic new normal. The inertia of the past is simply too strong.

That's why going cold turkey on cigarette smoking is such an ineffective strategy in terms of long-lasting results. Studies have shown that people who successfully quit smoking give up the habit many times. By cutting back a little, a lot, or completely, they accumulate the experience of success. The success lasts only a short time in most cases because of the physical side of tobacco addiction. Yet with repetition, the body does adapt.

Any significant change involves repetition. Developing new pathways in the brain is like digging a new river channel. Water will keep running down the old channel as long as it's deeper than the new one. By repeating the change you want to attain, you will be "digging" a shallow channel at first, but repetition deepens it. A physical metaphor can go only so far, however. Mental events are sometimes stronger than any physical history inside the brain.

People addicted to alcohol and tobacco sometimes kick their habit overnight, once and for all. The percentage of such people may be tiny (and overnight success isn't our aim in this book), but they remind us that the mind comes first in choice making, the body second.

This would be an arguable point to many biologists, who firmly believe that physical processes tell the whole story. But there's no need for argument, thanks to the intimate connection between mind and body. Every message you send to your body elicits a response, and the response will influence your next message. This circular dialogue, or feedback loop, is crucial. The choice to send new messages affects the entire feedback system.

2. Bad choices sometimes feel good.

The answer to using this principle to your advantage is to welcome gratification instead of judging it negatively. Are you shocked to hear that? To quote a phrase from the famous TV science fiction show *Star Trek: The Next Generation*, "Resistance is futile." Impulses and cravings have power over us because they hit in the moment. The brain opens a fast track to the desirable sensation, and the power of the rational mind to override the impulse gets postponed. However, studies have shown that a short pause is often enough to remedy this imbalance between reason and sensation. If a group of people wait for five minutes before acting on a craving, most of them won't give in. They find reasons not to, and the reasons suffice because the moment of instant gratification has passed. (There are even food lockboxes that come with a time-delay mechanism. Let's say you crave potato chips. When the craving strikes, you eat one chip and lock the rest of the bag into the box. It keeps the potato chips out of reach for a set time, typically between five and ten minutes, after which the lock releases. The idea sounds clever, but one wonders how many people are capable of eating just one potato chip when the craving arises, or who don't have other salty snacks ready and waiting in the cupboard.)

Instead of trying to manipulate your cravings, let go of the struggle. Look for instant gratification from better sources. The nutritionist's advice to eat a carrot instead of half a pint of chocolate gelato isn't realistic, but perhaps two Oreos will do the trick, or half a cupcake. There are few strategies that stop cravings and none that bring them to an end permanently, not by direct assault. The best approach is to reset your microbiome by instituting easy lifestyle changes and then rely on your body to return to a state in which it has no cravings.

There's also a major emotional component to cravings and the need for instant gratification. Dealing with this component successfully involves expanded awareness. When you discover what you're really hungry for, the answer will be something deeper than peanut butter and jelly or pepperoni pizza. As we'll discuss later, in the section on emotions, being fulfilled is an internal state that you can achieve if you know how to do so. Once you reach this state, the allure of external triggers will greatly diminish and then vanish. A craving for anything "out there" is best answered from "in here."

3. The gratification from good choices is usually delayed. The answer to working with this principle is that your microbiome can shorten the delay in gratification that usually follows good choices. The microbiome is constantly changing, and it responds quickly to diet, exercise, meditation, and stress reduction. As you continue to make small, easy good choices that also let you feel good right away, the positive effect of these choices begins to build. Very soon, instead of seeking to feel better, you will instead be trying not to lose the good feeling you already have. In contrast, someone addicted to instant gratification through making bad choices receives short jolts of pleasure that decrease over time, and only by feeding the craving is there any pleasure at all. Distraction from pain becomes the whole game.

By showing you how to work with the three big principles underlying choice making, we've put you in the position to create your own path to success. Being completely unique, you shouldn't be expected to follow a set regimen, whether it's the newest miracle diet, fat-burning gym routine, or power supplement. These methods all rely on the expectation that you will give up after a while and move on to the next profitable fad. What works is not restless wandering from one short-term solution to the next. Instead, you need to build a pyramid of easy choices that bring long-term results. The foundation of the pyramid is made out of the choices you consider the easiest to make. You then build the pyramid upward, level by level, with harder choices that have become easy thanks to the foundation. The capstone is radical well-being, which looks high and far away when you're standing on the ground but is almost effortless to achieve if you know what you're building and how to do it.

MAKING IT REAL

Let's give an example of pyramid building that actually comes from someone quite close to one of the authors. We'll call him Rudy's older cousin Vincent, although that's not his real identity. Vincent has been a practicing physician since the early eighties and has earned a name for himself in internal medicine. As often happens with doctors, Vincent doesn't practice what he preaches. His daily routine involves long hours without physical activity and with much exposure to the stress of hearing his patients' distressed reaction to illness. He prides himself on handling this very well. Years of dedication and ambition have made him what he is today, but Vincent has paid the price.

If he had come to himself as a patient, he'd be alarmed. Vincent carries forty pounds of excess weight. He drinks alcohol every day, sometimes to excess. He complains of insomnia and feeling fatigued. Recently the situation couldn't be ignored any longer because he developed joint pain, particularly in his knees. Undergoing surgical replacement only partially relieved his knee pain. You might think that the accumulation of these negative effects would have set Vincent, given all his professional knowledge, on the road to change, but that's not how human nature works. Having chosen denial as his chief tactic for dealing with his problems, Vincent had little choice but to double up on his denial as matters grew worse.

Then he made a discovery that got his full attention: the microbiome. Buoyed by the data, Vincent had found a way to bypass his denial while at the same time altering his lifelong view that only drugs and surgery are "real" medicine. The changes he made in his daily routine were all easy for him:

- Eating foods with soluble fiber like whole-grain bread, brown rice, bananas, oatmeal, and orange juice. This took care of his prebiotics, the food that intestinal bacteria feed on.
- Adding probiotic foods, which contain beneficial bacteria that would colonize in his intestines, primarily the colon. Active yogurt, sauerkraut, and pickles belong in the probiotic camp.
- Taking an aspirin a day for its anti-inflammatory effect.
- Cutting back on excess alcohol while not giving up his five-o'clock cocktail.

Vincent felt good about these easy changes, and he noticed results immediately in better sleep, pain reduction, and a general sense of feeling lighter.

He became convinced, as more and more doctors are, that fighting inflammation was the key. Now that he felt better, he regained his old optimism and hope. Getting rid of his problems seemed possible for the first time in years. The next stage of changes was made easy by his new attitude.

- He gave up drinking altogether. This wasn't a hard choice, because he was feeling so much better that he didn't need alcohol—and its inflammatory effects—as self-medication. At the same time he gave up the occasional cigar he used to enjoy with colleagues. The toxicity of tobacco became all too obvious to his palate and nose once they became sensitive again. Giving up smoking happened naturally as one outcome of his improved diet.
- He switched entirely to whole organic foods. There was no longer any attraction to foods with additives and preservatives, which were also possibly inflammatory.
- He reduced his salt intake, a craving that snack and junk foods heavily enforce. This was easy because his whole-foods diet had removed the desire to snack.
- After researching the possible benefits of taking a probiotic supplement, he chose one, with the intention of improving the kinds of bacteria that populate his microbiome.

Instead of suffering from a cascade of symptoms, many of them tied to inflammation and toxins leaking through the intestinal wall, Vincent was experiencing a cascade of recovery. Each easy step led to others that he would have considered hard choices if they had existed on a laundry list of good things to do. Instead, his lifestyle evolved day by day, and each change naturally led to the next.

Presently Vincent finds himself poised to make changes that were practically inconceivable even two months ago. Never a believer in the mind-body connection, he's now willing to take up meditation. The studies on the benefits of meditation have been around for decades, but only today does he make a personal connection with them—he's started thinking in terms of epigenetics and the microbiome, which are both affected positively by meditation.

After years of dependency on painkillers and drug treatment for his high blood pressure, Vincent has decided to wean himself off both. The first to go were the meds for hypertension, because a whole-foods diet reset his microbiome, and this was enough to regularize his blood pressure. The theme of countering inflammation, which was his original inspiration, has obviously paid off and may be leading to long-term benefits that aren't yet visible.

Your personal story—and your path to well-being—won't be the same as Vincent's. It shouldn't be. There is no one-size-fits-all, not when it comes to making choices that you can actually abide by. What will make your path similar to Vincent's is tending to the three issues in choice making. He applied the same answers being offered to you.

To overcome the problem of hard choices, Vincent made only easy ones every step of the way. Some of these would have seemed too hard at the outset, but they weren't once he had laid the proper foundation.

To overcome the problem of instant gratification, he stopped resisting his impulses, which went a long way to ending his guilt and self-judgment. He gave alternative gratification a chance through foods he enjoyed, and he trusted that alcohol and tobacco would fall away naturally, which they did after his chronic pain subsided.

To overcome the problem of delayed results, he made choices in which the results came quickly, primarily by changing to a whole-foods diet. Staying on the program didn't require patience and promises. You have to be patient if your choices don't alter the situation in your body until years afterward, as is true for anyone taking cholesterol-lowering drugs, for example—the heart attack they are trying to prevent lies years in the future (not to mention that such drugs may lower heart attack rates for a large enough sample of people but aren't guaranteed to prevent any specific heart attack, meaning yours).

You've probably noticed some areas that Vincent didn't bring into his new choices. The most obvious is exercise. He cherishes weekend golf games, which for now satisfy what he wants from exercise. But he also knows that golf isn't a cardiovascular activity, the kind of exercise that raises your heart rate and improves oxygen consumption, with the attendant benefits to cardiovascular function and blood pressure. Excess weight and joint pain had prevented him from doing this kind of exercise for a long time, so for Vincent, cardiovascular exercise still falls into the category of hard choices—a category that's always up for revision if you approach it with the attitude of building a pyramid one easy choice at a time.

Now you're prepared to construct your own pyramid, with each stone being *one new choice per week* that is easy to make. There are six categories of change that will have a meaningful effect on your epigenome, microbiome, and brain:

Diet
Stress
Exercise
Meditation
Sleep
Emotions

For each of these we'll offer you a menu of choices. Each menu will be long enough to present choices that are easy for anyone to adopt. Once you have circled your preferences in all six categories, you'll be ready to implement them with zero effort and every expectation of positive results. Pyramid building is the key to successful change that is lasting and cumulative.

Making changes one at a time selected from six different areas of your life increases their effect on the entire mind-body system. We recommend keeping track of the effects of your lifestyle changes by using the following list:

RESULTS TO LOOK FOR

Put a check beside each result that you begin to notice after adopting a new lifestyle change.

___ Digestion improves.

___ Upset stomach and/or heartburn decreases.

___ Constipation or diarrhea is no longer a problem.

___ Your body feels lighter.

___ You feel a growing sense of inner peace and calm.

___ Your thinking is sharper, more alert.

___ You are losing weight without dieting.

___ Signs of aging slow down.

___ Signs of aging actually reverse—you feel younger.

___ Life seems less stressful, and you can handle stress better.

___ Moods even out, no longer going up and down.

___ You have a sense of pleasant well-being.

___ Minor aches and pains lessen or vanish.

___ Hunger pangs lessen or vanish.

___ A natural cycle of hunger and satiation returns.

___ Headaches decrease or go away.

___ Bad breath lessens or vanishes.

___ Sleep becomes regular and uninterrupted.

___ Allergies improve.

___ Snacking is no longer a temptation.

___ Excessive sugar is no longer a temptation.

___ Cravings for addictive flavors (sweet, sour, salty) lessen.

___ Alcohol consumption decreases.

___ Tobacco consumption decreases.

For your doctor to verify:

___ Lower blood pressure

___ Normal blood sugar levels

___ Normal heart rate

___ Improvement in anxiety or depression, if present

___ Increase in HDLs (high-density lipoproteins, or good cholesterol)

___ Reduction in LDLs (low-density lipoproteins, or bad cholesterol)

___ Improved triglycerides (lowered risk of heart disease and stroke)

___ Normal kidney function

___ Better dental checkups (reduced plaques, cavities, gum inflammation)

DIET

Getting Rid of Inflammation

It won't come as a surprise by now that the biggest enemy in people's diet is inflammation. Medical researchers have tracked its footprints all over the map, from chronic disease and obesity to leaky gut syndrome and mental illness. The typical American diet is very likely to increase inflammation, and therefore a change is called for. The change will be drastic for anyone who subsists on junk and fast food. Yet the overload of sugar that enters into almost any diet if you aren't vigilant is also a prime suspect. Evolution didn't prepare us to consume over one hundred pounds of refined white sugar a year; it's not clear we evolved to consume it at all, along with the cheaper corn syrup that manufactured foods increasingly contain.

Inflammation is necessary to the healing process, when the immune system rushes chemicals known as free radicals to flood the wounded or diseased area. Almost all the symptoms of flu, such as fever and aches and pains, are not from the flu virus but from your body's recovery efforts and the inflammation that comes along with it. In this way, inflammation is our friend. Yet our friend can turn on us without our being aware.

You can be in a state of chronic inflammation without knowing it, because unlike the red, swollen areas that appear on your skin when it's inflamed, the internal signs of inflammation often go unnoticed. There is typically no feeling attached when the immune system is mildly compromised, and some signs of inflammation, such as joint pain, could have other causes. Our approach is to make easy choices that have an anti-inflammatory effect. An anti-inflammatory diet will cause most people to notice benefits right away.

Reading the menu: The menu of choices is divided into three parts, according to level of difficulty and proven effectiveness.

PART 1: EASY CHOICES

First are the choices that anyone can implement. If you begin to adopt them, you will be laying the foundation of your pyramid. As tempting as it is to adopt more than one easy choice at a time, resist the urge. Over the course of a year, you will be making fifty-two weekly changes in your lifestyle. There's no need to pile up on yourself.

PART 2: HARDER CHOICES

These are choices that you feel resistance about adopting, or which you know are too difficult to maintain without backsliding. That's perfectly okay. Harder choices can wait until you feel you have made all the easy choices you can. For some people the harder choices will actually be easy, because everyone has a different starting point. For most people, however, the harder choices are for higher up on the pyramid. They need to feel easy before you tackle them; otherwise you risk making a change that you can't continue.

PART 3: EXPERIMENTAL CHOICES

These are steps that have strong advocacy and intriguing research behind them, but which definitely constitute a minority position for now. Dietary fads come and go. Today's research gets modified or overturned tomorrow. Before adopting an experimental choice, read our caveats, pursue your own investigations, and make an informed choice. In any case, none of these experimental choices should substitute for the choices in Parts 1 and 2.

Remember that whatever choices you make are meant to be permanent. Since you are making only one change a week, you have seven days to see how it works out. If everything goes smoothly, you're ready to select a second change in the following week. Don't rush; don't put pressure on yourself. The secret to this strategy is making sure that it progresses effortlessly.

We think it's prudent to make dietary changes first, because food has the most direct effect on the microbiome. Our advice is to spend the first month entirely on dietary changes, but it's up to you. Before making any change, be sure that you've read all six sections of the program.

Diet: The Menu of Choices

Circle two to five changes that would be easy to make in your current diet. The harder choices should follow after you have adopted your easy choices, one per week.

PART 1: EASY CHOICES

- Add prebiotics with soluble fiber to your breakfast (e.g., oatmeal, pulpy orange juice, bran cereal, bananas, a fruit smoothie made from unpeeled fruits).
- Eat a side salad with lunch or dinner (preferably both).
- Add anti-inflammatory foods to your diet (see page 127).
- Consume probiotic foods once a day (e.g., active yogurt, kefir, pickles, sauerkraut, kimchi).
- Switch to whole-grain bread and cereals.
- Eat fatty fish at least twice a week (e.g., fresh salmon, mackerel, tuna, and canned or fresh sardines).

- Reduce alcohol to one beer or glass of wine a day, taken with a meal.
- Take a daily probiotic supplement and a multivitamin pill. Also take half an adult aspirin or one baby aspirin—see page 125.
- Reduce snacking by eating only one measured portion in a bowl—don't eat from the bag.
- Share dessert in a restaurant.

PART 2: HARDER CHOICES

- Switch to organic foods, including chicken and meat from animals not raised on hormones.
- Limit or eliminate red meat from your diet; at least switch to organic alternatives, including chicken and meat from animals not raised on hormones.
- Switch to "pastured" eggs high in omega-3 fatty acids (see page 143).
- Become a vegetarian.
- Cut out refined white sugar.
- Drastically reduce packaged foods.
- Eliminate alcohol.
- Stop eating fast foods.
- Stop buying processed foods.
- Stop eating when you're not hungry.

PART 3: EXPERIMENTAL CHOICES

- Adopt a gluten-free diet.
- Become a vegan.
- Eliminate wheat entirely.
- Have only fruit and/or cheese instead of dessert.
- Adopt a Mediterranean diet (see page 129).

EXPLAINING THE CHOICES

We won't need to explain every choice on the list individually, because there's a shared goal behind everything: fighting inflammation. In the easy category, your goal is to find effortless ways to combat inflammation. Chief among these is resetting your microbiome, where the digestive process starts the pathway that leads to inflammation. As we saw earlier, toxins produced by your gut microbes are safe as long as they remain in the digestive tract. But "leaky gut syndrome," which seems to be much more prevalent than previously expected, sends toxins into the bloodstream, and from there, the body fights the toxins using inflammation—a healthy response, but a dangerous one. Resetting your microbiome is the best defense and the first step to keeping these toxins where they naturally belong

Modern life exposes us to many influences that either harm the microbiome or are suspected to harm it, including the widespread use of antibiotics, a high-fat, high-sugar diet, lack of fiber, air pollution, excessive stress, bad sleep, and various additives and hormones in the food we buy. The microbes that colonize the gut are a direct cause of inflammation but also a protection against it when the microbiome is healthy.

You aren't aiming for a "perfect" microbiome, because no one can define such a thing, not yet at least. With over one thousand species of bacteria to consider, and with the microbiome being in a constant state of flux, perfection may be unattainable, or even the wrong thing to pursue. It's easier and more sensible to change your diet away from inflammation. There's no harm in doing so, and doing so offers the promise of many benefits.

Prebiotics come first. These are foods for the microbiome, chiefly from fiber that our own bodies can't digest. Evolution has led to a happy partnership in which bacteria consume the fuel they need without robbing our bodies of any, and vice versa. Prebiotic foods also buffer the body from inflammation by reducing endotoxin, a poison created by certain bacteria that is harmless inside the GI tract but highly inflammatory if it leaks into the bloodstream and activates the immune system. (See page 88 regarding the research that shows how a glass of freshly squeezed orange juice completely offsets the inflammatory effect of a high-fat McDonald's breakfast.)

Prebiotic foods aren't scarce. We recommend a breakfast that's rich in them, from bananas and pulpy orange juice to oatmeal, whole-grain breakfast cereal, and fruit smoothies made with unpeeled apple, various berries, and other fruits. You'll find countless recipes online, and the smoothie can be made with vegetables instead of fruit if that's what you prefer. Just be aware that green vegetables, the main ingredient in veggie smoothies, are much lower in calories than fruit. You don't want to eat a breakfast that's lower than 350 to 500 calories if you want enough energy to get to lunch without hunger pains and with sufficient energy. A salad with lunch or dinner also serves as a good prebiotic buffer.

Probiotics are foods that contain active bacteria. Active yogurt is the most common one at the supermarket, but there's also pickles, sauerkraut, kimchi (a traditional Korean fermented cabbage dish), and kefir (a fermented milk drink that tastes similar to yogurt). Including one of these foods during a meal helps to reset your microbiome by introducing beneficial bacteria that will colonize the walls of the intestine and hopefully reduce or drive out harmful bacteria. Because of the complexity of the microbiome and the huge differences from one person to another, there is no completely reliable prediction on the effects of probiotic foods. The best thing is to try them—all are completely harmless—and then look for results.

Probiotic supplements are a booming business that's expected to rise dramatically in the future. Health food stores offer a bewildering variety of these supplements, some in pill form to be taken on a full stomach, others in perishable form that must be refrigerated. There is no expert medical advice about the best probiotic supplements, for the same reason that crops up repeatedly: the microbiome is too complex and is constantly shifting. It should also be noted that a reliable supplement that contains 1 billion bacteria will enter a gut ecology of 100 trillion microbes. Outnumbered 100,000 to 1, the supplement may have negligible impact.

We prefer to be optimistic. Any opportunity to reset the microbiome to a state of natural balance is worth taking. A supplement can't substitute in any significant way for getting your probiotics through food, yet it's an easy choice to take a supplement. Also, to augment the benefit, add a multivitamin and one baby aspirin, or half an adult aspirin, to your routine. The aspirin is a proven way to reduce the risk of heart attack and some kinds of cancer. (Be sure to consult your doctor before combining aspirin with other drugs, particularly those that have anti-inflammatory or blood-thinning properties.) The multivitamin isn't a must if you are eating a balanced diet, but as we age, the intestinal tract becomes less efficient at processing vitamins and minerals. Studies have shown that up to one-third of dementia cases are linked to mineral deficiencies or poor diet.

Dementia is a generic term that covers a host of disease conditions, including Alzheimer's disease, which Rudy studies, and there's no accepted dietary regimen that is guaranteed to be preventive. But research that focuses on how food affects brain cells has come up with a few general guidelines that are easy to follow; most are directly in line with an anti-inflammation diet. The preventives are

Omega-3 fatty acids found in fatty fish (For those who are alarmed by heavy metals present in fish oil, an alternative source is organic flaxseed oil along with a handful of walnuts every day. If you do choose fish oil, use triple-distilled oil to avoid heavy-metal contaminants.)

Antioxidant micronutrients (blueberries, dark chocolate, green tea) to fight free-radical damage in the brain

B vitamins (not more than the recommended daily allowance)
A Mediterranean diet (see page 129)

Keep in mind that these are provisional suggestions. Even a supplement like vitamin E, which has been promoted for decades for its antioxidant effects, has run into contrary research. The basic neuroscience revolves around the fact that brain tissue is quite vulnerable to free-radical damage, because the brain uses 20 percent of the total oxygen consumed by the body. Free radicals are molecules with an extra oxygen atom that is quick to find another molecule to bind with. Although necessary for healing wounds as part of the whole inflammatory response, free radicals in excess can damage healthy cells through unwanted chemical reactions; brain cells seem to be a prime target in cases of dementia.

Reducing potential damage from overly active oxygenation is the common link connecting most of the preventives listed above, but totally validated proof is lacking. Our position is that a balanced diet is the best way to protect yourself, but taking a supplement may be helpful, particularly if you are over age sixty-five. A common effect of aging is reduced kidney function, which often results from low-level inflammation of the kidneys, or nephritis. Decreased kidney function diminishes the body's retention of the water-soluble vitamins B and C. Taking a multivitamin supplement, then, makes sense if you are older. The main drawback for most people is that vitamins don't usually have any discernible benefit that you can feel, and the damage that can be traced to inflammation, including excess free radicals, should be addressed directly through an anti-inflammatory dietary regimen.

Anti-inflammatory foods have come into favor with increasing public interest and research studies. If you are primarily interested in seeing a list of specific anti-inflammatory foods, you can find a generally agreed-upon list at www.health.com. But it is much more effective to understand the whole issue of inflammation, because a holistic approach attacks the problem from many angles instead of just one. The following foods are listed primarily to reinforce your knowledge, not to tell you that only these "right" foods belong in your diet.

Foods That Fight Inflammation

Fatty fish (but see the caveat about heavy metals on page 125)
Berries
Tree nuts
Seeds
Whole grains
Dark leafy greens
Soy (including soy milk and tofu)
Tempeh
Mycoprotein (from mushrooms and other fungi)
Low-fat dairy products
Peppers (e.g., bell peppers, various chilies—the hot taste isn't an indication of inflammatory effects in the body)
Tomatoes
Beets
Tart cherries
Ginger and turmeric
Garlic
Olive oil

In their online health publications, Harvard Medical School adds a few items to the list:

Cocoa and dark chocolate
Basil and many other herbs
Black pepper
Alcohol in moderation (but also see page 133)

Other listings add the following:

Cruciferous vegetables (cabbage, bok choy, broccoli, cauliflower)
Avocado
Hot sauce
Curry powder
Carrots
Organic turkey breast (substitute for red meats)
Turnips
Zucchini
Cucumber

Needless to say, these are all healthy whole foods, and making them a mainstay of our diet can only be beneficial. However, the science is still out on whether all of these foods have an anti-inflammatory effect in the body, and also what effect, if any, they have on the genome, epigenome, and microbiome. Still, the fact that your super genome responds to every experience strongly suggests that what you eat has consequences at the genetic level. The fact that so many diseases are connected to bad diet proves that there's a genetic connection, so our best advice is that a good diet is one way to promote better genetic activity.

On the opposite side, there are also foods that increase inflammation, as listed by the same bulletin from Harvard Medical School.

Foods to Limit or Avoid

Red meat
Saturated and trans fats (e.g., animal fats and the hydrogenated vegetable fats found in many processed foods)
White bread
White rice

French fries
Sugary sodas

To these, other reliable sources add

White sugar and corn syrup (frequently hidden in processed foods that aren't primarily sweet)
Omega-6 fatty acids (see page 140)
Monosodium glutamate (MSG)
Gluten (see page 134)

Our feeling is that an anti-inflammatory diet has to be better than an inflammatory one, because the foods that are proven risks—junk food, fast food, fatty and sugary foods—also lead to inflammation. The link between inflammation and chronic disease is too strong to ignore, and paying attention has many benefits.

The Mediterranean diet has a good reputation for being healthy. A 2014 study conducted in Spain made headlines by proving with statistical accuracy that subjects who ate a Mediterranean diet lowered their risk of heart attack considerably. In fact, the results were so positive that the study was cut short, since it became unethical to allow the other subjects to continue on their non-Mediterranean diet. There have been no similar studies of an anti-inflammation diet (in fact, the Spanish study was the first of its kind to be conducted with such scientific rigor), but the overlap is significant. A Mediterranean diet replaces red meat with fish, and butter with olive oil. Alternatively, for vegetarians like Rudy, noninflammatory protein can be obtained from other sources like tempeh, tofu, and mycoprotein (e.g., Quorn and Gardein products). Whole fruits, vegetables, low-fat tree nuts (e.g., almonds and walnuts), and seeds (e.g., chia, hemp, sunflower, pumpkin, flax) are also recommended. When you add all of these up, you'll see that some of the most important anti-inflammation foods are there in the Mediterranean diet.

Why, then, do we place the Mediterranean diet under experimental choices? There are several reasons. First is the permanence of such a change. Sticking with the diet comes easily if you are a native of the region and have been on it since childhood, but the Mediterranean diet is not so easy as a lifetime choice if you are used to the typical Western diet. Also, unless you live alone, there's the issue of asking your family to make the change with you. But just as important is the science. The kind of study that was conducted in Spain is about risks as they pertain to large groups. It's a numbers game. Going on the Mediterranean diet doesn't guarantee that any individual is protected, while our aim here, to fight inflammation, is completely about the individual. Still, the Mediterranean diet comes close to being an anti-inflammation diet, so it's very worthwhile to try, but only after you've made other, easier choices to see if you've accomplished the same goal.

Switching to olive oil brings up the tangled issue of *fats in the diet*. Our primary advice is to avoid trans fats, chiefly hydrogenated oils found in packaged foods and at some, but not all, fast-food chains. These oils are known to have inflammatory effects. Limiting saturated fats in butter and cream and avoiding red meat also seem prudent.

You need to have a healthy balance of blood lipids (fats), including cholesterol and triglycerides. Both are necessary for cell building and repair. Blood lipids are processed by your liver after you ingest fat in your diet. This processing is quite complex, depending on diet, genes, weight, age, illness, and other factors. Problems can arise for anyone who is obese, whose liver is genetically predisposed to deliver too much cholesterol to the body, who suffers from a hormonal imbalance, or whose immune system has been activated by inflammation, among other factors. It's not as simple as "ingest more cholesterol, and your cholesterol levels will go up." To further cloud the issue, the leading drugs for lowering cholesterol, known as statins, do not seem to reduce the risk of heart attacks, according to studies going back to 2010. This indicates what has long been known, that heart attacks depend on more than just cholesterol.

We feel that inflammation, which is strongly linked to heart disease, is the first culprit to go after. The damage it causes can be traced back to the gut-inflammation connection. With so many risk factors tied to inflammation, it seems better and easier to work on it as a whole rather than singling out "good" and "bad" fats. We aren't endorsing saturated fats by any means. Polyunsaturated cooking oil, and especially olive oil, remains the healthiest choice.

Another issue is how much fat you should be eating. People find it quite difficult to cut back their fat consumption all at once, even though extreme fat restriction has long been part of the heart health program devised by Dr. Dean Ornish at the University of California, San Francisco. Ornish's lifestyle-driven approach to heart disease led to extraordinary results. His program of diet, exercise, meditation, and stress reduction remains the only proven way to reverse the plaque that lines the coronary arteries in people at high risk for heart attacks. Ornish also pioneered in studies showing that his program creates beneficial changes in the genome through epigenetic switching of hundreds, now thousands, of genes, a process known as upregulation.

To clear coronary arteries of plaque, as Ornish has accomplished, requires a severe cutback in fat intake, to as little as one tablespoon of added fat a day. The standard recommendation from the American Heart Association allows for fat to be 30 percent of one's daily calorie intake—a huge difference. (Even reaching 30 percent is difficult, considering that the average American diet, although around 34 percent fat, which doesn't seem far off the mark, has actually added an extra 340 calories a day over the past two decades. This amounts to a potential weight gain of over 30 pounds a year.)

We support and acknowledge Dr. Ornish for his invaluable work, but severe fat restriction leads to noncompliance. Cutting back to only a few tablespoons of all fats and oils a day, or as little as one tablespoon if you are being rigorous, simply taxes the average person too much. Low-fat diets for weight loss probably fail around 98 percent of the time, insofar as that's the failure rate for all crash diets. Our approach of building a pyramid of easy choices doesn't include severe fat restriction.

Besides noncompliance, we have another good reason, we believe, for not putting a strong emphasis on fats or on cutting back on calories as the road to *weight loss*. Animal studies strongly suggest that the microbiome may be the real key. As we alluded to earlier, simply by inserting microbes from obese mice into other mice with the same genome leads to weight gain in the normal mice. Anecdotal evidence from self-experimenters like Dr. Zhao in China leads to the same conclusion, as does the small study with identical twins in which one twin is obese and the other twin lean.

Resetting the microbiome through an anti-inflammation diet is win-win. It will either lead directly to weight loss or put you in a state of balance in which moderate calorie cutting becomes feasible without backsliding. We've summarized our weight-loss strategy in the following list.

Basic Steps for Weight Loss

- Don't follow a calorie-restricted diet. Leave calorie cutting for the end, not the beginning, of your weight-loss regimen.
- Focus on the easy steps for reducing inflammation first.
- Put your attention on prebiotic and probiotic foods.
- At the same time, make easy choices about increasing your physical activity. The most important step is to stop being sedentary and to move throughout the day.
- Attend to good sleep, since bad sleep throws off the key hormones for hunger and satiation.
- Make easy choices relating to emotions, since emotional eating is generally a component of weight gain.

- After following the above steps for at least 3 to 4 months, assess if you are losing weight. A loss of ½ pound per week would be considered a high benchmark. A loss of 2 pounds a month is still a success. If you've lost that much, keep doing what you're doing without cutting calories.
- If you see no weight loss, consider cutting 200 calories from your daily intake as long as that's easy for *you*. Consider this a permanent choice like the other easy choices on the program.
- If it's not easy to cut calories, keep making other changes and check back on your weight in 2 months. Reassess calorie cutting then.

Alcohol has had its medical proponents for a long time, and the public tends to accept that the French have lower heart attack rates because of the national habit of wine drinking. In the list of anti-inflammatory foods, the Harvard Medical School online site includes a drink a day (although not defined, this presumably means one beer or one glass of wine) because of a single beneficial effect: it seems to lower levels of C-reactive protein (CRP), a powerful signal of inflammation. More than one drink (the source of the alcohol doesn't appear to matter) increases CRP, however. In general, alcohol has been classed as inflammatory. It is metabolized very quickly, like refined white sugar, and we consider it in the same class as white sugar when it comes to potential damage throughout the system.

But we are also realists and realize that social drinking is deeply embedded in the West and is increasingly catching on in Asia. People don't like giving up something they enjoy. Therefore we offer an easy choice to limit yourself to one drink a day, preferably as part of a complete meal so that the metabolic rush of the alcohol is tempered by other food. Our hope is that by adopting easy changes that reset your microbiome and send positive messages to your epigenome and brain, you will no longer want to drink. You'll feel good enough without it, and your sense of well-being will actually be lessened by not having any alcohol at all.

Reducing *gluten* in your diet also falls under the experimental heading. The number of people considered by mainstream medicine to suffer from a gluten allergy is tiny (the most common diagnosis is among those with celiac disease, which severely damages the intestines), but there's a widespread belief, amounting to a crusade, that countless others are feeling the ill effects of gluten. As anyone soon discovers when trying to eliminate gluten from their diet, it appears in many processed foods, not just in the usual source that comes to mind, which is wheat and wheat products.

Symptoms of gluten sensitivity, often generalized as "wheat belly," include bloating, diarrhea or constipation, distended abdomen, and abdominal pain. This list, which centers on digestion, has been extended by some advocates to other symptoms elsewhere in the body, such as headache, generalized pain, and fatigue. Self-diagnosis is the most common route, because doctors look for specific allergic responses recognized as celiac disease or the most typical alternative, nonceliac gluten sensitivity. Medical training also pinpoints various disorders, like irritable bowel syndrome, that have much the same range of symptoms; or wheat allergy, which is sometimes present without sensitivity to other sources of gluten.

Since we are asking you to make easy choices first and foremost, going on a totally gluten-free diet isn't one of them. The list of foods you would have to give up is long (provided by www.healthline.com):

Bread, pasta, and baked goods made from wheat (or wheat bran, wheat germ, or wheat starch)
Couscous
Cracked wheat
Durum
Farina

Farro
Fu (common in Asian foods)
Gliadin
Graham flour
Kamut
Matzo
Semolina

Wheat isn't the only grain that contains gluten, so you'd also need to cut out

Barley
Bulgur
Oats (oats themselves don't contain gluten but are often processed in plants that produce gluten-containing grains and therefore may be contaminated)
Rye
Seitan
Triticale and mir (hybrids of wheat and rye)
Veggie burgers (if not specified gluten free)

Gluten may also show up as ingredients in barley malt, chicken broth, malt vinegar, some salad dressings, and soy sauce, as well as in many common seasonings and spice mixes. A gluten-free diet requires total dedication. For the sake of completeness, we'll list the grains that are permitted on such a diet.

Amaranth
Arrowroot
Buckwheat
Cassava
Millet
Quinoa

Rice
Sorghum
Soy
Tapioca

Of course you also have the choice of limiting gluten-containing foods rather than eliminating them entirely. Both of us have been intrigued enough to try eliminating gluten in our own diets, and we are very enthusiastic about the results in increased energy, balanced appetite, and some weight loss. It should be realized, however, that the scientific validation remains to be seen for "wheat belly" as a widespread ill and for wheat sensitivity as a problem affecting millions of people.

If you are still intrigued, go ahead and try an experiment for a week. A simple diet of rice instead of wheat is the foundation for billions of Asians. You would also cut out pasta and the vast majority of baked goods. But this isn't onerous now that gluten-free sweets are on the market, and you don't need to resort to them if you have nonprocessed sweets like flan or bake with gluten-free flour. The results of our experiment are likely to be quite good, since an Asian diet minus pasta, bread, cake, pie, and cookies is already quite healthy, leaving aside the controversial issue of gluten sensitivity.

Vegetarian diets have long been considered a healthy alternative. We have made a personal choice to move toward a plant-based diet. Rudy has been vegetarian since his university days, but when faced with a busy schedule, he does consume some dairy for the purpose of quick protein. In India, the Brahmin, or priest, caste traditionally subsists on a meat-free diet, and for many people excluding meat is a humanitarian measure connected with the killing of animals. For most people, however, vegetarianism represents a hard choice. Being naturally high in fiber, a vegetarian diet is very likely to be anti-inflammatory and beneficial to the microbiome as well. Why, then, aren't lifelong vegetarians free of chronic disease?

Actually, many are. The current data show that vegetarians are at lower risk for

Heart disease
Colorectal, ovarian, and breast cancer
Diabetes
Obesity
Hypertension

These findings don't sort out the anti-inflammation factor, so there's no way of knowing the status of vegetarians who also avoid refined sugar, alcohol, high stress, and a sedentary lifestyle. Until there's a study on people who have adopted a holistic lifestyle aimed at reducing inflammation, being a vegetarian stands as a very good choice if it's easy for you, but it's by no means a panacea.

On a comparative scale, it's much easier to follow a vegetarian diet than a *vegan diet.* Like a vegetarian diet, a vegan diet is plant based and excludes meat, but it also typically excludes all dairy products (milk, cream, yogurt, butter, cheese), along with eggs and all products that contain these ingredients. A strict vegan diet therefore involves a meticulous regimen for getting adequate protein. Soy (in tofu or tempeh) is a complete protein and as such is generally a major source of protein for many vegans, and also for vegetarians.

Your body needs nine amino acids, the building blocks of protein, that it cannot produce itself. It's not necessary to have all of them at every meal, and for vegetarians, a diverse mixture of vegetables, fruits, seeds, and nuts will be sufficient. However, there are some foods for vegetarians in addition to soy that contain all nine of these essential amino acids, including quinoa, buckwheat, hemp seed, chia, and the simple food combination of rice and beans.

Rudy limits his soy intake to one meal a week so as not to overload on phytoestrogens—naturally occurring compounds in soy that are similar to human estrogen. Although current research tends to show that males don't suffer a risk of lower testosterone from phytoestrogens, Rudy has made this personal choice in terms of his intake of hormones.

Besides these protein sources, to make sure you're getting adequate protein as a vegan you would use combinations of foods that contain various amino acids, the building blocks of protein, to get a full complement—that is, a complete protein. (The usual route is to combine legumes, grains, potatoes, and even mycoprotein, e.g., in Quorn products, in different combinations.) We have put vegetarianism under harder choices and vegan diet under experimental choices for the above reasons. Having been a vegetarian since college, Rudy, along with his entire family, fully enjoys this lifestyle choice.

THE SCIENCE BEHIND THE CHANGES

Both the epigenome and the microbiome play crucial parts in how food affects your body at a much deeper level than was ever suspected. When nutritionist Victor Lindlahr titled his 1942 book *You Are What You Eat*, he did more than coin a popular phrase; he foresaw by decades the research that would support the diet-gene connection. Now numerous studies exist, primarily with mice, showing that diet is in fact the main factor influencing the composition of the microbial genome we harbor in our gut. For example, switching suddenly from a vegan to an animal-based diet changes the microbiome in just days. In a study at the University of California, San Francisco, mice were fed either a high-animal-fat, high-sugar (junk food) diet or a low-fat, plant-based (vegan) diet. When the animals were then switched from the vegan to the junk food diet, the host of intestinal microbes (as assessed in their feces) changed within three days regardless of the genetics of the mice used. Diet mattered much more than genes. This finding helps to explain why identical twins with identical genomes can have as many differences in their microbiome as two siblings who aren't twins and therefore have similar but not identical genomes.

Diet also dramatically affects epigenetics, as we saw earlier in the example of the Dutch famine during World War II. In rural Gambia, for example, there is a rainy (hungry) season, when nutrition is low in protein and energy, and a dry (harvest) season, when the diet is heavy in vegetables and high-energy foods. Children of 84 mothers conceived during the hungry season had lower birth weights and higher levels of epigenetic modifications (methylation) in their genome than those conceived by 83 mothers during the harvest season. (There were also major differences in B vitamin and folic acid levels in maternal blood samples in the two seasons, which correlated with the epigenetic changes.)

The children born to mothers who experienced an unhealthy diet during conception were also more likely to go on to develop insulin resistance and type 2 diabetes. Naturally, these facts underscore the need for all pregnant women to maintain a healthy diet, but the larger point was expressed nearly two centuries ago when the noted French gastronome Jean Anthelme Brillat-Savarin wrote: *"Dis-moi ce que tu manges, je te dirai ce que tu es"*—"Tell me what you eat, and I will tell you what you are."

MAKING THE SCIENCE WORK

When people look for information about diets, three forces pull at them. All three are supposedly based on science, yet they contradict one another.

First is the standard nutritional advice to eat a balanced diet. This advice changes slowly. It is well established in nutritional studies. The problem is that people don't comply. In the face of solid science, the American diet continues to move in the wrong direction (i.e., high fat, high sugar, an overload of calories, reliance on junk and fast foods).

Second is cutting-edge research studies. These studies can be very intriguing, and the studies on inflammation and the diet represent a major breakthrough. The problem is lack of human trials on a wide basis, along with findings that contradict each other.

Third is the latest fad diet for weight loss. These diets typically make overstated claims and seem to change every day, using "breakthrough" research that may be flimsy or badly skewed. Sometimes no real science exists in support of the latest diet. Yet the public rushes to follow the latest fad until a new one is touted on the grapevine.

We've taken a stand on some cutting-edge research despite the absence of large-scale human trials. Countering inflammation, as with the Mediterranean diet, seems scientifically sound to us. In any event, an anti-inflammation diet overlaps with standard nutrition in almost every area and thus provides a second source of scientific validation. However, there are areas of confusion in an anti-inflammation diet that should be faced honestly.

Fatty acids are a prime example of such an area of confusion. There has been rising awareness that the omega-3 fatty acids found in fatty fish are good for you, and standard nutrition advises everyone to eat such fish once or twice a week. Yet there's another group of fatty acids, omega-6s, that complicate the story. Your body needs both omega-3s and omega-6s, and because it can't make them, they must come through diet. What makes these substances special is that unlike other fats, the omega group isn't used primarily for energy, but for biological processes, including the production of red blood cells.

It seems to be crucial, according to various studies, to keep omega-6 levels down, because high levels are strongly linked to inflammation. Improvements in heart disease and rheumatoid arthritis have been shown by getting the balance between omega-3s and omega-6s back into the healthy range. All Western diets are too high in omega-6s because of the heavy use of polyunsaturated cooking oils. Yet these oils, made from vegetable sources—corn, soy, safflower, and so on—were once considered the healthiest ones, with risk factors for heart attack as the primary support for this claim.

Today the evidence has strongly moved in another direction. Studies of indigenous peoples (who use few processed vegetable oils and eat no processed packaged foods) indicate that the ratio of omega-6s to omega-3s in their diet is about 4:1. In contrast, Western diets are fifteen to forty times too high in omega-6 foods, with an average ratio of omega-6s to omega-3s of 16:1. At such high levels, the omega-6 fatty acids block the benefits of the omega-3s. Genetic studies aren't easy to come by in this area, but it's speculated that we evolved in hunter-gatherer societies to consume a diet even lower in omega-6s, with a ratio of omega-6s to omega-3s closer to 2:1. In the body, getting closer to a 1:1 ratio seems ideal, according to some experts.

Among foods high in omega-6s, cooking oil leads the way, but there are others, as follows:

Main Sources of Omega-6 Fatty Acids

Processed vegetable oils—highest are sunflower, corn, soy, and cottonseed
Processed foods using soy oil
Grain-fed beef
"Factory-raised" chicken and pork
Non-free-range eggs
Fatty cuts of conventionally raised meats

As you can see, the polyunsaturated oils that are a major part of standard disease prevention turn out to have a serious drawback in terms of inflammation. The only vegetable oil that is low in omega-6s and high in omega-3s is flaxseed oil. Safflower, canola, and olive oil aren't particularly high in omega-3s but are the lowest in omega-6s among commonly sold vegetable oils, with olive oil the best.

Adding to the confusion, "bad" saturated fats like lard, butter, palm oil, and coconut oil are low in omega-6s. This is one reason why standard nutritional advice has begun to recommend a balance of saturated and polyunsaturated fats. But the real culprit, it seems, isn't so much the food we eat in its natural state but processed foods. Soy oil is cheap and readily available, lending itself to use in hundreds of packaged foods. Beef raised in feedlots on grain to achieve maximum bulk in the shortest period of time are much higher in omega-6s than grass-fed beef (not to mention the widespread use of antibiotics and hormones in the beef and dairy industry). Also high in omega-6s are pork and chicken produced on conventional grain feed in the "factory" system, along with factory eggs.

This is why one of the harder choices we present is switching to grass-fed beef, along with naturally fed (also called pastured) chickens and their eggs. "Free range" isn't always reliable, since the birds might still be receiving some conventional feed. What makes this choice hard is that it's expensive, and most of the sources aren't supermarkets.

We didn't bring up the issue of omega-6 imbalance to alarm you, only to illustrate the complexity of how food interacts with the body. Rebalancing the fatty acids in your diet comes down to some easy steps, with a general emphasis, as mentioned previously, of moving toward a plant-based diet, as the authors have chosen to do, even though it's not strictly vegetarian:

Restoring Fatty Acid Balance

- Cook with safflower and olive oil; canola oil isn't as good but is acceptable.
- Eat unsalted or low-salt tree nuts, including walnuts, almonds, pecans, and Brazil nuts. Limit the amount of

fatty nuts, such as cashews and macadamias, as well as peanuts.
- Eat seeds, including unsalted chia, sunflower, pumpkin, hemp, and flaxseeds.
- Eat fatty fish—no more than 6 ounces per week. If vegetarian, eat more lower-fat tree nuts, such as walnuts and almonds, and seeds.
- Avoid packaged foods with soy oil high in the list of ingredients.
- Don't cook with soy, sunflower, or corn oil.
- Cut back or eliminate conventionally raised beef, pork, and chicken.
- With any meat and poultry, buy lean cuts and trim the fat from other cuts.

There is indication that our diets shouldn't just be lower in omega-6s but much higher in omega-3s. Therefore it's a major challenge to turn the American diet around. (Vegetarians who rely heavily on soy products like tofu and whole soybeans would be even more challenged.) Should you make a major push toward omega-3 fatty acids? Some experts believe that these should actually outnumber omega-6s in the diet, but we think the jury is still out on this issue. Among native populations, the Inuit, with their traditional marine diet and high intake of fish, are the only ones who have reversed the ratio, with omega-3s outnumbering omega-6s by 4:1. The Inuit were held up in the early excitement over omega-3s as examples of people with a very low risk of heart disease. But later studies found that the evidence for this claim was fragile, and in addition, the blood-thinning properties of omega-3 fatty acids may be why Inuit have higher than normal mortality from strokes. The larger point is that getting excited over "miracle" foods or nutrients and worried over forbidden ones is a recipe for confusion. The great strength of human digestion is its adaptability. We are the ultimate omnivores. But we are also the only creatures who modify their diet according to ideas in our head and the traditions we are born into.

We respect innovative ideas and traditions, but they also can be excuses for resisting good science and pursuing fads. Taking the route of easy choices seems best. The story doesn't end with diet, of course. There are five more areas of lifestyle that complement the ability of food to change your microbiome, epigenome, and brain activity. Sometimes they work through anti-inflammation, yet there are other mechanisms that bring major benefits, too. Easy choices with life-changing results can come from many directions.

STRESS

An Enemy in Hiding

Being told to reduce the stress in our lives is largely falling on deaf ears. Modern life *is* stress. There's no escape from the external pressures (technically known as stressors) that make everyone's existence too fast, too exhausting, and too demanding. Asking people to have less stress is like asking fish to have less water. We can try to shrug off stress as normal because it's so prevalent, but the body cannot. Even an experience that might seem totally positive, like winning the lottery or going on vacation, can trigger the same stress hormones as negative events.

Most people accept that stress is harmful, excepting highly competitive types who claim to thrive on stress. An adrenaline junkie might rush to free-climb a rock face without ropes, skydive, or wrestle an alligator with the full backing of media coverage that extols the rush of a thrill-seeking life. But medical science disagrees. The surge of stress hormones—principally adrenaline and cortisol, which carries the stress response forward—can be interpreted as a thrill. Hidden from sight is physiological reality. These hormones lead to a cascade of reactions, including elevated heart rate and blood pressure that your body is meant to endure for only a brief period under acute conditions. When prolonged and repeated, the stress response starts to damage tissues and organs throughout the body.

The hidden danger is from chronic stress, which is so constant and low lying that we fool ourselves into believing we've adapted to it. The body tells a different story. Imagine the following image:

A shell-shocked soldier has been brought home from the battlefront. He looks numbed and dazed. He complains of exhaustion but cannot sleep. Sharp, sudden noises create a state of alarm. When he's not agitated, he is mentally dull and quite often depressed.

This is the classic picture of acute stress when it has been prolonged beyond the ability of the body to recover properly. Shell shock was thought once to be a sign of weakness or cowardice, but now we know that its basis is physiological. Despite the fact that our tolerance for stress, like the tolerance for pain, varies widely from person to person, all soldiers will succumb to shell shock if subjected to acute stress hour after hour, as happened to combat troops under constant shelling in the trenches during World War I.

Now imagine yourself sitting down to watch TV in the evening when suddenly the dog next door starts to bark. You attempt to put the noise out of your mind, but the dog won't stop. This doesn't count as acute stress. You won't jump up with a classic fight-or-flight response. Even so, you are being subjected to the same three factors that aggravate all stress.

Repetition: The dog keeps barking and won't stop.

Unpredictability: The barking came out of nowhere, and you don't know when it will end.

Lack of control: You have no way to directly stop the dog from barking.

It's these three factors that generally lie behind the problem of chronic stress. Of course, they affect a frontline combatant far more severely. Being shelled repeatedly, at unpredictable times, and without being able to stop the enemy artillery, multiplies the actual danger a dozen times over as compared with that from a neighbor's barking dog. The stress response exists to protect you from danger, however, and despite the higher brain's ability to tell the difference between a barking dog and trench warfare, the lower brain is stuck millions of years ago in evolutionary time. It signals stress hormones to be secreted by the endocrine system, not in a flood but on rheostat control, as it were. The drip-drip of a low-level stress response is as destructive as Chinese water torture, and for the same reason. Given enough tiny, harmless stresses, the path to total breakdown is open.

Everyone's goal should be to prevent the aggravating factors in stress. We consider this true stress management. In the menu of choices below, many stressors can't be entirely eliminated; modern life simply doesn't allow for it. But there are important ways to improve your body's reactions by inserting better messages into the feedback loop. After discussing the choices and what they mean, we'll revisit the science that applies to stress management.

Reading the menu: As in every section on lifestyle, the menu of choices is divided into three parts, according to level of difficulty and proven effectiveness.

Part 1: Easy choices
Part 2: Harder choices
Part 3: Experimental choices

Please consult page 120 in the diet section if you need a refresher on what the three levels of choice are about. Remember, too, that whatever choices you make are meant to be permanent.

The question immediately arises, should you double up on choices, one for diet and one for stress? We know that for some people there's a sense of urgency to make changes in more than one area, and if you see easy choices in two areas—not just diet and stress but any two of the six lifestyle areas we cover—it's your choice to adopt them at the same time. We don't think it's the best strategy, though. If you overlap two choices, it's more likely that you will lapse. Permanent change depends on making things easy and absorbing any new change into your existing lifestyle. One at a time seems like enough. Remember, if you change only one thing a week, that's fifty-two changes a year, which represents an enormous shift.

You'll immediately notice that meditation is the first choice listed under stress. There's a complete section on meditation beginning on page 171, which is where the main discussion takes place. To us, meditation is the most important strategy for reducing the stress response and rebalancing the mind-body system. Keep this in mind even though there are many other easy choices here. In the list of harder choices, we advise dealing with negative emotions. That discussion can be found in the emotions section beginning on page 201, but we consider it a main source of protection from stress, too.

Stress: The Menu of Choices

Circle two to five changes that would be easy to make in your current stress management. The harder choices should follow after you have adopted the easy choices, one per week.

PART 1: EASY CHOICES

- Meditate every day (see page 171).
- Decrease background noise and distractions at work.
- Avoid multitasking. Deal with one thing at a time.
- Stop being the cause of someone else's stress (see page 151).
- Vary your daily activity, including time out and downtime (see page 152).
- Leave work on time at least three times a week.
- Stop unloading your stress on family and friends.
- Avoid people who are sources of pressure and conflict.
- Be in contact with people who are meaningful to you.
- Decrease boring and repetitive work.

- Reduce alcohol to one beer or glass of wine a day, taken with a meal.
- Take up a hobby.
- Retreat from stressful situations quickly.
- Find a physical outlet to unwind from daily stress.

PART 2: HARDER CHOICES

- Seek the most meaningful work you can find.
- Be a manager instead of a worker.
- Seek job security over money.
- Save for the future. Be fully insured.
- Become more accepting.
- As much as you can, stop resisting.
- Stop taking on too much responsibility.
- Stop bringing work home. Leave the office at the office.
- Take more days off from work.
- Eliminate boring, repetitive work.
- Enjoy Nature every day.
- Find a close confidant.
- Find a mentor.
- Adopt a vision of the future.
- Become a healer of stress (see page 157).
- Deal with your negative emotions—anger, fear, anxiety, self-judgment, depression (see page 201).

PART 3: EXPERIMENTAL CHOICES

- Become your own boss.
- Work toward a secure sense of self and higher self-esteem.
- Become someone's close confidant.
- Become a mentor.
- Take a course in crisis management.

- Deal with long-standing psychological issues through therapy.

EXPLAINING THE CHOICES

We've already mentioned that learning to meditate, which is a prime strategy for stress management, will be dealt with in a section of its own. Otherwise, you'll notice that we've focused on work and the workplace. We've done this for two reasons. First, almost everyone has to work with other people in an atmosphere where stress inevitably arises; second, the other main source of stress, relationships, would need a whole book of its own, given how different all families are. Making changes at work will teach you how the general principles apply, and any reduction in stress can't help but have benefits at home.

Staying with the workplace for now, everyday pressures fall into three categories: time pressure, peer pressure, and the pressure to perform. It's rare for anyone to be without these pressures as long as work means deadlines, coworkers, and performance goals. So how do you adapt to these constants? Most people are reactive. They pay little heed to their repetitive patterns of behavior; therefore, they are highly ineffective at dealing with stress.

Bad Ways to Cope with Stress

How many of the following ineffectual ways do you use to deal with daily pressures at work?

I react emotionally and sometimes blow up.
I complain about the pressure I'm under, mostly to people who aren't causing it.
I pass the stress down the line, unloading it on someone else.

I turn my back on the people who cause me the most stress, blocking them out as much as I can.
I put up with stress until I get a chance to unwind (e.g., going to the gym, cocktail hour).
I put even more pressure on myself and others, on the theory that it makes me stronger and more competitive.

These behaviors are generally unconscious, because when examined rationally, they don't achieve what they set out to do—decrease the harmful effects of stress. Stress is a feedback loop. The input is the stressor (e.g., a tight deadline, an obnoxious boss, an unreachable sales goal); the output is your response. You have a choice to intervene anywhere along the loop by changing the input or the output. The more consciously you intervene, the higher your chances of reducing the bad effects of stress.

In our menu of easy choices, some are directed at input, some at output. For example, you can stop multitasking, which has been shown in brain studies to decrease performance and increase inattention. You can perhaps reduce external noise and distractions around you at work. Both changes are on the input side. On the output side, you can improve your response to stress—you can stop passing your stress on to other people, for example, and walk away from stressful situations as soon as possible.

Perhaps the most important easy change, however, is to *stop being the cause of stress* for others. This involves more self-awareness than the other easy choices, and becoming more self-aware is the closest thing we know of that can be called a panacea, or cure-all. Some of the bad ways to manage stress have already been mentioned. Basically, they involve putting stress on others when you should be coping with it yourself. Many of us do this inadvertently by bottling it up inside, closing down the lines of communication that could solve the problem. Going to the gym to unwind may be good for you, but it does nothing for the atmosphere at work. A high-strung boss only makes for stressed employees.

You are the source of stress when you make a habit of complaining and criticizing. Complainers also find it hard to praise and appreciate others. You are creating stress when you indulge in perfectionism, never being satisfied until every *t* is crossed. Even normal office behavior like forming cliques and gossiping behind someone else's back is, if we face reality, a source of stress that can be emotionally devastating. It amounts at times to little more than bullying, a source of stress that goes without saying. Hold your behavior up to the mirror, then see page 157 to find out how you can become a healer of stress instead. As you begin to see the results of being more self-aware, you can turn to the harder choices on the menu, which mostly deal with deeper habits that aren't easy to break.

Time management can also reduce stress in ways most people don't pay attention to. Varying your activity throughout the day opens many possibilities. Office work is sedentary, and the human body is meant to move. Getting up from your chair once an hour is enough to reverse some of the adverse effects of a nonphysical job. Decades ago a Yale physiologist took student athletes and had them lie in bed without getting up for an extended period—staying in bed was the traditional protocol for hospital patients recovering from surgery as well as new mothers. After two weeks in bed, the athletes lost the equivalent of two years of training as their muscles wasted away. Unexpectedly, it wasn't simply staying in bed that caused the damage—gravity played a part. If the subjects stood up during the day, even doing minimal activity, most of the muscle wasting didn't occur, which is one reason that postoperative and maternity care now emphasize getting up and around as soon as possible.

Besides standing up and moving around at least once an hour, you should make space during the workday for downtime, when you simply relax, and inner time, when you either meditate or simply sit quietly with your eyes closed. These activities allow the whole system to reset itself. In addition, you will feel more centered psychologically. Taking a few simple steps, it turns out, counters the tendency for repetitive work to dull the mind. It's the kind of low-level stress that often goes unnoticed.

The harder choices are self-explanatory, except for one: becoming a manager instead of a worker. According to an old punch line, the boss says, "I don't get heart attacks. I give them." There's physiological truth in that. The more independent you are, the less you find yourself following orders from above, the lower your stress level. This finding is unrelated to how many hours you work. The higher up you rise on the corporate ladder, the more likely you will love your job but the more likely you will also take it home with you. People who love their jobs typically report that they work eighty hours a week between the office and home.

Only the CEO of a company reports to no one higher up (he loses sleep over what the shareholders are demanding), which brings up one of our experimental choices. These choices focus on getting more independence by starting your own business, which most of us see as the ideal. But independence means more than being your own boss. Developing a long-range vision for your life offers a much more meaningful kind of independence. Working on your deep psychological issues opens the possibility for psychological freedom, making you independent of your past and the scars you carry around with you. These are meaningful choices that go beyond the limited definition of stress management, yet it's this kind of change that transforms someone's life.

THE SCIENCE BEHIND THE CHANGES

Stress was the first area where the mind-body connection could be proved, opening the door for the flood of research and validation that exists today. The main reason for focusing on stress was probably simplicity. It's exacting and difficult work to extract a neurotransmitter like serotonin or dopamine from brain tissue. You must work with samples from dead tissue instead of in real time, and of course the subjects are rarely human. But stress hormones like cortisol and adrenaline rush into the bloodstream in real time and can be sampled on the spot by drawing blood. In addition, the physical effects of fight-or-flight are easily observed in ourselves.

Significant findings fine-tuned what was happening, which is how stress researchers were able to prove that unpredictability, repetition, and lack of control are the aggravating factors in stress. In a classic experiment, mice were placed in cages wired to administer mild electrical shocks. In itself, each shock was harmless. But the experiments administered the shocks repeatedly at random intervals, and the mice had nowhere to escape. After only a few days, the animals became dull and listless. Their immune response was severely compromised, and some actually died from the "harmless" shocks. This experiment allowed for the understanding of how low-level chronic stress damages the body. It also dispelled the myth that succumbing to repeated stress was a sign of weakness or some other character flaw—the physiology simply can't help itself.

In the era of epigenetics these findings have penetrated to the deepest level of our physiology, with increasing hope that people can modify and improve their stress response. Not only the food you eat but also your level of stress can cause epigenetic modifications and alter gene activities. In a study of the effects of the Holocaust on gene activity, researchers at Mount Sinai's Icahn School of Medicine in New York City took 80 children who had at least one parent who was a Holocaust survivor and compared them with 15 "demographically similar" children whose parents didn't go through the Holocaust. The results are described in a moving first-person account by one child of a survivor, Josie Glausiusz, in a June 2014 issue of *Nature*.

For two weeks in the spring of 1945, Glausiusz's father, "his mother and three surviving brothers had been packed onto a train along with 2,500 other prisoners of Bergen-Belsen, the concentration camp in Germany where my father had been incarcerated since 6 December 1944," she writes. "For 14 days, while the family survived on minuscule rations of scavenged raw potato peels and maize, the 'Lost Train' snaked haphazardly through Eastern Germany, blocked by the advances of the Russian and American armies, before halting in a forest near the small German town of Tröbitz."

Unknown to the passengers trapped in boxcars, their German captors had decoupled the locomotive and escaped during the night. Suddenly two Russian cavalrymen appeared on white horses and systematically broke the locks that held the camp prisoners in.

Having grown up on this harrowing tale, Glausiusz volunteered for the Mount Sinai study in 2012. It was led by Rachel Yehuda, a neuroscientist and the director of the school's division on traumatic stress. The aim of the study was "to determine whether the risk of mental illness owing to trauma is biologically passed from one generation to the next. In particular, the researchers wanted to see whether such risk could be inherited through epigenetic marks."

Reporting on what her participation entailed, Glausiusz writes, "During the course of the study, I completed an online questionnaire to assess my emotional health as the daughter of Holocaust survivors and whether my parents had post-traumatic stress disorder (PTSD). A psychologist interviewed me about my parents' war-time experiences and my own history of depression and anxiety. I submitted to blood and urine tests measuring the hormone cortisol, which enables the body to respond to stress, as well as the methylation of GR-1F, a promoter of a gene that encodes a glucocorticoid receptor, which binds cortisol and helps shut down the stress response."

The findings turned out to be somewhat contradictory, depending on which parent suffered from PTSD as a Holocaust survivor. To simplify, the key was determining if epigenetic marks led to more or less cortisol circulating in the bloodstream of their children. Children whose parents both had PTSD were found to have more gene activity leading to the production of the glucocorticoid receptor that helps turn off the stress response by binding with cortisol (i.e., rendering it ineffective). Turning on the gene turns off the stress.

The results were mixed with a single parent who had PTSD. It appears "that children of fathers with PTSD are 'probably more prone to depression or chronic stress responses,' Yehuda says. [But] the reverse seems to occur in the offspring of mothers with PTSD." These children showed lower cortisol levels. Why?

One possible explanation: "Mothers who survived the Holocaust, [Yehuda] says, often feared separation from their children. 'When you've been exposed to a lot of loss, and you're very worried that you will keep losing loved ones, you may literally hang on too tight.' Holocaust offspring, she says, often complain that their mothers were overattached to them.

"Although she does not identify the mechanism behind these changes, Yehuda thinks that epigenetic modifications might occur before conception in the fathers, but that in the mothers the changes occur either before conception or during gestation."

We were reluctant to bring up such horrific experiences, except that this Holocaust study marked a breakthrough. According to Yehuda, as far as her team was aware, "This is the first evidence in humans . . . of an epigenetic mark in an offspring based on preconception exposure in a parent." (A previous experiment in mice, already mentioned, had shown that whether a baby mouse had good or bad mothering led to epigenetic marks that affected the stress response; nurturing behavior by the good mothers reduced anxious behavior in their offspring along with lowered cortisol levels.) It's also important to note that the study is controversial, largely because the biochemistry of gender differences is complex, and the differences found by Yehuda were small, or as she puts it, "nuanced." It should also be noted that without being able to spot the epigenetics involved, psychiatry had long been aware, through various studies, that the effects of PTSD can be passed on to children of Holocaust survivors.

MAKING THE SCIENCE WORK

An old joke says, "Gray hair is inherited. You get it from your children." The science shows that it turns both ways. We may care a great deal more about how stress will get passed on in our families than at work. But the best approach in both places is the same: become a healer of stress. Your behavior today is likely to have consequences far into the future.

When you have an awareness that you aren't just the victim of stress but a potential source, your behavior changes. Here are some positive choices to relieve the stress around you at work, and they can be applied to relationships and family as well.

How to Be a Healer of Stress

How many of the following positive behaviors do you practice?

Asking others how they feel and listening to the answer.
Not insisting that you get your own way.
Always showing respect for everyone. Never belittling or scapegoating.
Never criticizing someone in public.
Accepting input from as many people as possible.
Praising and appreciating other people's work.
Being loyal in order to win loyalty.
Not gossiping or backbiting.
Waiting until you are calm before addressing a situation that makes you angry.
Giving coworkers and employees enough space to make their own decisions.

Being open to new ideas, no matter whom they come from.
Not favoring a small circle to the exclusion of everyone else.
Addressing tension as it arises instead of denying it or
hoping it will solve itself.
Not being a perfectionist who can never be satisfied.
Treating both sexes equally.

If you have already adopted most or all of the behaviors listed here, congratulations—you are already a healer of stress. Most of us, however, must make a conscious effort to change our ways, either in small or large part. None of us are being subjected to lab experiments on stress, yet in a very real way our lives are the laboratories in which we confront a host of stresses. It's up to us to become self-aware so that we understand the part we play in a world all but overwhelmed by demands, pressure, and crises. The individual is the source of healing, a truth that never wears out with retelling.

EXERCISE

Turning Good Intentions into Action

The secret to exercise can be told in a single phrase: keep going, don't stop. It's better to be active all your life at any level, including mild activity, than to play sports in high school and college, only to sit back as the years advance. Consistency is the main goal, not breaking a sweat. But this takes a conscious choice, one you are willing to stick with. The good news is that the more you keep on moving your body, the more you'll want to. Physical activity becomes a habit you adapt to rather quickly, not to mention that it helps create new pathways in the brain.

Modern life has made exercise a blessing and a curse. The blessing is that we are no longer slaves to backbreaking physical labor; the curse is that the blessing has gone too far. Modern life for most people is physically too soft, yet despite the price our bodies pay, we seem to prefer it that way. Given a choice, most people choose

Sitting still instead of moving around
Pleasurable distractions (TV, video games, the Internet) instead of playing sports
Mental work instead of physical work
Letting machines instead of muscles perform physical tasks

Letting their children spend more time on the computer and less time playing outside

These are all modern choices, and the trend hasn't stopped moving in their direction. As long as it does, the drawbacks of a sedentary life, such as increased obesity and type 2 diabetes, will plague society, while the benefits of exercise—in terms of cardiovascular health, avoidance of some types of cancers, and improved mental status—will be missed opportunities. As of 2013, only 20 percent of American adults got the recommended amount of regular exercise, which is 2.5 hours of moderate aerobic exercise per week or half that time spent in vigorous aerobic exercise. Someone between the ages of eighteen and twenty-four is twice as likely to exercise as someone over sixty-five—31 percent versus 16 percent—even though it's evident that the two groups that benefit the most from physical activity are the very young and the very old.

For our ancestors, rest was a luxury; for most of us, finding the time to go to the gym is the luxury. At the turn of the twentieth century, around 80 percent of the calories expended to run a farm still came from the farmer using his muscles. This was true despite the invention of farm machinery and the widespread use of horses to draw plows, harvesters, and wagons. Such a life, where physical activity was hard and constant, was how we evolved. Our bodies are well adapted to much more activity than you'd suppose. There is evidence that primitive hunter-gatherers had a life span as long as seventy years. What shortened their lives were external conditions—disease, childhood mortality, exposure to the elements—not the built-in frailty of the body.

Because most of us don't have to hunt, gather, till the soil, fork hay into the hayloft, or make our own bread—the list can be extended ad infinitum—there's almost no essential physical work left. Therefore no matter how often we hear the drumbeat of diet and exercise, good intentions outweigh action. It's because compliance is so low that we put stress management above exercise in our lifestyle list. More people are more likely to reduce the pressure in their daily lives than to get up out of their chairs and start moving.

We are realists, and we know that scolding will never motivate people to change their ways. Guilt only leads to unused gym memberships. Neither will the balance of pain and pleasure serve as motivation. Anyone who enjoys exercise is highly likely to have been running, lifting weights, or playing sports since childhood. Their bodies are conditioned to it, and the feedback loop that leads to the runner's high or to the "good tired" of a workout is a source of pleasure. For someone who isn't in the habit of exercising, though, the reverse is true. Exercise affects the body like physical labor, leading (at the beginning) to fatigue and sore muscles. The body of someone who doesn't exercise is habituated to sitting still, the ill effects of which are mostly long term. It can take years before the reality of heart disease, type 2 diabetes, and excessive weight actually begins to dawn.

Our goal, then, is to provide easy choices that can change the feedback loop, which means that a little activity leads to wanting more. In addition, the recommended changes must be maintained for a lifetime. Getting active in spurts with long periods of no activity in between isn't good for you. Adaptation comes naturally when it's regular and steady. Better to walk up a flight of stairs every day than to shovel snow off the driveway six times a winter.

Reading the menu: As in every section on lifestyle, the menu of choices is divided into three parts, according to level of difficulty and proven effectiveness.

Part 1: Easy choices
Part 2: Harder choices
Part 3: Experimental choices

Please consult page 120 in the diet section if you need a refresher on what the three levels of choice are about. You should make one change per week total, not one from each lifestyle section. Remember, too, that whatever choices you make are meant to be permanent.

Exercise: The Menu of Choices

Circle two to five changes that would be easy to make in your current level of physical activity. The harder choices should follow after you have adopted the easy choices, one per week.

PART 1: EASY CHOICES

- Get up and move around once an hour.
- When taking an elevator, take the stairs to the second floor before pressing the button.
- Do your own housework instead of hiring a cleaner.
- Take a brisk walk after dinner.
- Choose the far corner of a parking lot (as long as it's safe and well lit).
- If you already walk your dog every day, make the walk longer and brisker.
- If a destination is less than half a mile, walk instead of drive.
- Buy an exercise step and use it for 15 minutes every day as you watch TV or listen to music.
- Go outside for 5 to 10 minutes three times a day.
- Take up gardening, golf, or a similar activity that you actually enjoy.
- Set aside 5 to 10 minutes a day for calisthenics.
- Do more than half the chores around your house.
- Work with light weights as you watch TV.

PART 2: HARDER CHOICES

- Acquire more-active friends and join them in their activities.
- Devote half of your lunch hour to exercise.
- If you take children to the park, play with them instead of watching.
- When using an elevator, take the stairs to the third or fourth floor before pressing the button.
- Plan a shared exercise activity with your partner or spouse twice a week.
- Buy an exercise step and use it for at least 30 minutes every day as you watch TV or listen to music.
- Resume a sport you used to love.
- Do 5 to 10 minutes of calisthenics twice a day.
- Walk for a total of 3 hours a week.
- Do all your own yard work.
- Volunteer to help the needy with housecleaning, painting, and repairs.
- Take hikes every weekend in good weather.
- Use a trainer at the gym.

PART 3: EXPERIMENTAL CHOICES

- Join an exercise class.
- Take up yoga (see page 165).
- Lead a hiking group.
- Train for a competitive sport and keep at it.
- Find a regular exercise buddy.
- Take up tennis.

EXPLAINING THE CHOICES

The easy choices on the menu are quite easy. They would have to accumulate quite a lot to equal the official recommendation of 2.5

hours of moderate aerobic activity a week, combined with some additional time at weight training. But those recommendations might as well come from another planet if you lead an inactive life. The good news is this: Getting up out of your chair brings the most benefit. Moving away from a completely sedentary life is the major step in preventing the bad effects of getting no exercise. The risk of disease rises sharply as you age if you don't move around. Drastic inactivity eventually leads to a 30 percent higher mortality rate for men and double the mortality rate for women. The "new old age," in which seniors remain active and vital well beyond sixty-five, reversed one of the unhealthiest trends in social life.

The more activity you add, the better your body will respond. If you go from jogging a mile to running a mile, the good effects will increase. What your heart, brain, circulatory system, blood fats, and blood sugar need most is *some* activity, after which you can think about adding more.

In middle age, getting physical decreases the risk of chronic illness. Statistical measurement has proved the point over and over. Unlike other risk factors, however, exercising is more than statistical. It improves every individual life, at every level of activity. In very old people, eighty and above, weight training for a few minutes with minimal effort (using only a five-pound weight, for example) can double or triple muscle tone.

Our focus isn't on how much weight you can lift or how fast you can run. We want to level the curve so that physical activity isn't mostly for the young, with a sharp falling off in middle and old age. Leveling the curve is much more important than being really active in your youth and inactive in old age. Your body adapts to what you do *all the time*, not what you do every once in a while. This is also the secret for making exercise pleasurable—the feedback loop between muscles and brain gets enlivened the more you use it. Just like a biceps or abdominal muscle that atrophies with disuse, the body's feedback loops need to be utilized, and the more messages they transmit, the livelier they become.

Of course, we hope that you will move on to the harder choices on the menu. Give it time. If you spend two months taking the stairs to the second floor before pushing the elevator button, the next step—walking to the third or fourth floor—becomes effortless. But if you decide tomorrow to walk to the fourth floor, you are likely to feel exhausted, and your body will get the message "This is work." It's not the right message, not if you intend to make taking the stairs a pleasurable choice.

If we had to choose the single activity that does the most for body and mind together, it would be *yoga*. The correct term is *Hatha Yoga,* which is only one limb of the ancient tradition of Yoga, which has eight limbs in all. The others have to do with mind and behavior, but the body cannot be excluded in the pursuit of higher consciousness. In Sanskrit *Yoga* means "union" and is related to the English word *yoke.* As mysterious as the concept of enlightenment may seem, Yoga makes sense in its goal of bringing the mind, body, and spirit into harmony. Each position (or *Asana* pose) that's taught in Yoga is about focusing the mind to direct the flow of physical energy in the body.

Not that the two are separate. When consciousness moves, so does the energy. The teachings of Hatha Yoga can be quite subtle and even esoteric. The flow of life energy (*Prana*) that is regulated by the breath can be trained in exquisitely precise ways. The flow of life energy connected directly by the mind (*Shakti*) is even more precise and exact. It's taught that a single syllable in a mantra, for example, has influences that extend from mind and body throughout the entire environment.

The topic is so fascinating that we are devoting a section to consciousness as the pivot between everyday well-being and radical well-being. Hatha Yoga is a step in that direction. It improves body awareness, gets you back into physicality, sharpens your focus, and tones your muscles at the same time. Ironically, the practice is taken up mostly by men in India and mostly by women in this country. In India, the pursuit of higher consciousness is open to everyone in theory, but in practice women have been excluded. In America, men typically disdain yoga because it's not weight training or aerobic. Both attitudes are skewed and need to change.

THE SCIENCE BEHIND THE CHANGES

At the moment, the epigenetics of exercise is so new that few studies exist, but this hasn't stopped genetics from making its biggest contribution. We now know that being holistic isn't just someone's personal preference—it's necessary for everyone. Because hundreds and sometimes thousands of gene activities are changed through lifestyle choices, exercise can't be isolated from diet or diet from stress. This shift has enormous implications.

For example, health care providers used to minimize the health risks of leading a sedentary life. If you asked a physician thirty years ago what was wrong with having no physical activity, almost the only thing he'd come up with is disuse atrophy—the wasting of muscle tissue when a muscle isn't being used. Now we realize that a broad range of mind-body problems arise from a sedentary lifestyle, spanning heart disease, anxiety and depression, hypertension, and diabetes. The affectionate image of a plump grandma sitting in her rocking chair has become an image of bad health and decreased well-being.

These ill effects can be gleaned by looking at statistics for the general population, but epigenetics will one day be able to fine-tune an individual's personal risk. Sometimes what's true for a large number of people isn't true for you the individual. Across the population it's a well-established fact, for instance, that inactivity leads to obesity through the simple formula that expending fewer calories than you take in will develop body fat. But as we've seen, the old belief "calories in, calories out" has been revised.

To get at a possible genetic link between physical activity and body fat, a study carried out at Lund University in Sweden investigated the effects of physical activity on epigenetic gene modifications in fat cells. The researchers found that exercise led to epigenetic shifts in gene activity (via methyl marks) that affected fat storage in the body. They looked at the genomes of fat cells in twenty-three healthy men aged thirty-five before and after attending aerobics classes for six months, roughly twice a week. They found that exercise led to epigenetic changes in seven thousand genes, many of which led to genome-wide changes in DNA methylation in fat cells, shifting activity to enhance fat cell metabolism.

Methylation can remove methyl groups if they are properly exposed by histones, which work hand in hand with DNA in epigenetic modification by either exposing it to epigenetic marks or burying it—in essence, the switch is either made available or not. With exercise, methylation patterns change: some genes are silenced by methyl marks and others are unsilenced by demethylation. These are complex changes, but in essence, switches are turned off (downregulated) for pro-inflammatory genes while anti-inflammatory genes are switched on (upregulated). No doubt the mounting evidence about lifestyle changes will expand the anti-inflammation story across the entire mind-body system.

Weight loss is a common goal when people begin to exercise, but exercise leads to mixed results. The number of calories consumed through physical activity isn't as great as people suppose. A slightly brisk walk burns 280 calories per hour. Hiking, gardening, dancing, and engaging in weight training burn around 350 calories per hour. At 290 calories per hour, bicycling under 10 miles an hour burns off little more energy than walking. If your physical activity is vigorous—running, swimming, or aerobics—energy consumption increases to between 475 and 550 calories per hour. But even playing a vigorous game of baseball burns off only 440 calories per hour. Considering that a medium-size blueberry muffin contains

425 calories, there's good reason why exercise alone isn't the solution to weight loss.

However, if we take a holistic perspective, so much else changes when you become physically active that calories diminish in importance. In one study, overweight people were divided into three groups. The first group ran a mile, the second group jogged a mile, and the third group walked a mile. At the end of the trial period, the group that lost the most weight was the one that walked. One reason is metabolic. Once you break into a sweat, your body goes from aerobic metabolism, which burns calories, to anaerobic metabolism, which doesn't. So there are instances in which less pain means more gain. Keeping exercise light but constant seems to be the key. Yet even this bright note is offset by the fact that exercise, being physical work, can make you hungrier. In addition, heavier exercise builds muscle mass, which is heavier than body fat. We've considered these variables and keep coming back to the basic principle that you should make easy changes and keep on going, not stopping.

Very little has been discovered about the epigenetic effect of trying to lose weight. On the one hand, it appears that adult obesity goes back to childhood and adolescent experiences that extend into later years. Methylation may have imprinted bad habits and overeating into a person's gene activity. There is also the question of how much of an epigenetic influence gets passed on from obese parents to their children. We've been citing the data from the Dutch famine of World War II, but that evidence comes from extreme starvation, which then led to genetic modifications that apparently raised the risk of obesity in children, depending on whether their mothers were pregnant during famine times or times of plenty. It's quite another thing to sort out epigenetic marks according to which cause is at work, since obese parents can easily pass on bad eating behavior as well as epigenetic marks from their own experience before and during pregnancy.

Just as significant may be a Spanish study that took 204 obese or overweight teenagers and put them on a ten-week weight-loss regimen. It's well known that being obese as an adolescent leads to higher risk for a range of diseases in adulthood, not just the risk for being an obese adult. The program in this study was multifaceted. The teenagers were given personalized diet and exercise programs. They attended weekly meetings that gave them more nutritional and exercise information, along with psychological support.

At the end of the ten weeks, the researchers selected out the subjects who were considered either high or low responders to the program, depending on BMI (body mass index, which looks at the percentage of fat in the body) and the amount of weight lost. Looking at their epigenomes, some strong correlations were found. The high and low responders showed differences in methylation in ninety-seven different sites along their DNA. As reported online at the epigenetics site EpiBeat, there was a link to inflammation. "The involved genes belong to networks related to cancer, inflammatory response, cell cycle, immune cell trafficking, hematological system development and function."

In five sites the changes were so different that simply by examining the methyl marks there, one could predict who would be a low or high responder to a weight-loss program. The differences increased, the better someone responded to the program. These results offer two possibilities. First, epigenetic profiling may enable us to know in advance who will find it easy or hard to lose weight. Second, we may be able to pinpoint the gene activities that physical exercise promotes.

Making the gene connection more precise solves only part of the problem. It was originally thought that methylation occurred in the womb and lasted for a lifetime. Now it's realized that epigenetic changes are dynamic, constant, and often very rapid, taking place in twenty-four hours. Chemicals known as demethylases can remove methyl marks, and they have been connected to a specific gene (for fat mass and obesity-related transcript). Variants of this one gene are more associated with risk for obesity than any other genes. As reported by epigenetic researchers at the University of Alabama, Birmingham, it's thought that the instructions encoded in FTO create a protein that acts as a demethylase. This protein may act to turn off or on the genes that create obesity, although the exact mechanism isn't known, nor is it known why FTO is related to obesity. But the key finding is that regular exercise "largely erases the increased risk for obesity associated with the versions of the FTO gene. No one then is doomed by their genes," said team leader Molly Bray.

When it comes to the microbiome, there's been little study connecting it directly with exercise. One intriguing finding, however, comes from Ireland, where a team from the University College Cork compared forty professional rugby players with a control group of healthy adult males. The athletes were at pre-season training camp, which is a controlled environment—they ate and played together. The investigators looked at blood markers for inflammation that were also connected to immunity and metabolism.

It turns out that the athletes had a much more diverse microbiome. They were also improved over the control group in regard to markers for inflammation, immune response, and metabolism. Although some of the improvement could have been through diet, this seems to be a significant finding, if a very general one, about how gut microbes respond to exercise.

Given the present state of the science, we feel that the best practical course is to rely on demethylation through positive lifestyle choices—in other words, doing what you can today to regulate the genes that are beneficial, with a focus on lowering markers of inflammation. To date, there's no way to target only the changes related to body weight, but that's not essential for most people who aren't significantly overweight. A general program of the kind we're recommending is the best medicine anyone has yet devised with good science behind it.

MEDITATION

The Centerpiece of Your Well-Being?

The title of this section poses a question. Should you take up meditation as the primary choice for improving your well-being? Its benefits are cumulative. The more you do it, the better the results. But how many people begin to meditate and then stop after a while? In our experience, this has become a bigger problem than convincing someone to start. The same frazzled pressures that motivate people to seek the quiet oasis of meditation also cause them to quit. The excuses are generally about not having enough time or simply forgetting to meditate. Many look upon meditating as a kind of Band-Aid for patching up an especially bad day. "I'm feeling good today. I don't need to meditate" goes along with the notion of meditation as a quick boost, like a protein shake.

Our focus in this section will be on why meditation should be a lifelong practice. We know this is a major lifestyle change. It represents a unique kind of commitment, and the inconvenience can be considerable. Stopping to meditate breaks up the active routine of the day; it isolates you from contact with other people; its benefits are largely invisible. For all that, devotion to meditation also brings unique benefits.

It's a modern twist to look upon meditation for physical results, but studies on blood pressure, heart rate, and stress-related symptoms were the opening wedge that brought meditation into public acceptance in the West. Having your doctor recommend meditation bypassed the issue of whether you should "believe in it." This was a huge divergence from the East, where meditation has traditionally been for enlightenment, a concept the West looked upon suspiciously as an unfathomable mystery and probably unattainable except by swamis, yogis, gurus, and mystics.

The same fork in the road still exists. As a lifestyle choice, meditation appeals to people who want to see improvements in their health. As a spiritual choice, meditation appeals to people who want to reach a higher state of consciousness. It's this second group, we strongly suspect, who keep meditating regularly for years and perhaps a lifetime. Their goal may be invisible, but it's clear and creates long-term motivation. On the other hand, if you take up meditation to feel better, there's not a strong reason to do it on the days when you already feel good.

MEDITATION AND SUCCESS

Our way to get past this problem is simple: Make meditation the centerpiece of your total well-being. Adopt it, not because you are motivated to meditate, but because you will use it as a means to get something you want very badly. Only a need that is tied to desire will be fulfilled. Desire is the most powerful motivator, but in most people's lives there's no need to meditate the way there's a need for food, shelter, companionship, money, and sex. One strong desire, however, is general enough and long-lasting enough to fit the bill: success. If meditation can be linked to success, we feel that many more people would stick with it.

Making this connection requires a major shift, however. Both sides of the meditation divide—those who want better health and those who want higher consciousness—focus on a goal that is very different from worldly success. If you listed the most prominent traits of millionaires, entrepreneurs, and CEOs of major corporations, their success wouldn't be attributable to meditating. But the stereotype of the ambitious, competitive, and ruthless climber doesn't square with reality.

The bottom line is that *success* is a more potent word—and a stronger motivator—than *prevention*, *wellness*, and *well-being*. The attributes of highly successful people can be linked with the benefits of meditation.

Elements of Success

The ability to make good decisions
A strong sense of self
Being able to focus and concentrate
Not being easily distracted
Immune to the approval or disapproval of others
Sufficient energy for long workdays
Not easily discouraged
Emotional resilience, bouncing back after failure and setbacks
Intuition and insight, being able to read a situation ahead of others
A stream of new ideas and solutions
A cool head in a crisis
Strong coping skills in the face of high stress

If these aren't yet considered the key traits associated with success, they should be. Each trait is strengthened through meditation. How many people realize that they can make better decisions if they meditate, or keep a cooler head in a crisis? The stereotype of the navel-gazing, self-absorbed meditator is just as false as the ruthless climber clawing his way to success. The main reason that meditation caught on for many people in the West was that doctors and psychologists found a way around the image of the world-renouncing yogi with a long beard isolated in his Himalayan cave. But only recently has research in altered gene activities proved that meditation creates thousands of changes with holistic implications for mind and body.

That's a great advance, but attitudes need to shift even more. When success is defined by externals—money, possessions, status, and power—it's granted to the few, who usually begin from a privileged background. But what if success is defined differently, as an inner state of fulfillment? If you turn within, you can be successful at this very moment, because success is a creative process. You are engaged in it already, because true success is something we live. It's not an end state we arrive at. This is the message Deepak has been spreading for thirty years and exemplifying in his own life. It's the message he takes to business schools every year, teaches to CEOs, and expands upon in books like this one—and Rudy found that even before they met, he and Deepak had been walking the same path.

Reading the menu: As in every section on lifestyle, the menu of choices is divided into three parts, according to level of difficulty and proven effectiveness.

Part 1: Easy choices
Part 2: Harder choices
Part 3: Experimental choices

Please consult page 120 in the diet section if you need a refresher on what the three levels of choice are about. You should make one change per week total, not one from each lifestyle section. Remember, too, that whatever choices you make are meant to be permanent.

Meditation: The Menu of Choices

Circle two to five changes that would be easy to make in your current lifestyle in regard to meditation. The harder choices should follow after you have adopted the easy choices, one per week.

PART 1: EASY CHOICES

- Take 10 minutes at lunchtime to sit alone with eyes closed.
- Learn a simple breath meditation for use 10 minutes morning and evening (see page 176 for instructions).
- Use a mindfulness technique several times a day (see page 176 for instructions).
- Take up simple mantra meditation for 10 minutes twice a day (see page 176 for instructions).
- Find a friend to meditate with.
- Take inward time whenever you find it helpful, at least once a day.

PART 2: HARDER CHOICES

- Join an organized meditation course.
- Increase your meditation to 20 minutes twice a day.
- Make meditation a shared practice with your spouse or partner.
- Add some simple yoga poses to precede your meditation.
- Add 5 minutes of *Pranayama* (breath technique) before meditating (see page 177 for instructions).
- Teach your children to meditate.

PART 3: EXPERIMENTAL CHOICES

- Investigate the spiritual traditions behind meditation.
- Go on a meditation retreat.
- Become a meditation teacher.
- Explore taking meditation to the elderly.
- Explore introducing meditation at a local school.

EXPLAINING THE CHOICES

The easy choices on the menu are about finding minimal time during your day to go inward. The simplest means are a kind of premeditation, simply sitting with eyes closed or even defining "inner time" any way you want, just as long as you get to be alone with yourself, eliminating as much external noise and distraction as possible. Of course, we hope you are ready for meditation itself, yet if this is going to be a permanent change, don't rush into a commitment you can't keep. Fortunately, many people are surprised by how easily they take to meditating and enjoy the opportunity for inner time every day.

Breath meditation: This is a simple technique that takes advantage of the mind-body connection. Your breath is a fundamental bodily rhythm that is connected to heart rate, stress response, blood pressure, and many physiological rhythms. But it's also connected to mood—notice what a relief it is to take deep breaths when you are upset, and how ragged the breath becomes when you feel anxious or stressed. A breath meditation, then, helps restore the whole system and brings deep relaxation without effort.

The technique is simple. Sit with your eyes closed in a quiet place. Once you feel settled, follow your breathing as it goes in and out. Don't force your breathing into a rhythm or try to make it change. If your attention gets distracted by stray thoughts or sensations, easily bring it back to your breath. Some people find it helpful to put their attention on the tip of the nose, where the sensation of inhaling and exhaling is easy to focus on. Continue to follow your breath for the period you've set aside as meditation time, but sit and relax for a moment after you're finished. Don't jump up and become active immediately.

Mantra meditation: One of the most intricate and subtle branches of the Indian spiritual tradition has to do with sound (*Shubda*). The specific mantras that emerged from this tradition were valued for their vibrational effect, not their meaning. In the modern era there is no consensus about how thinking a specific word could affect the brain, and yet thousands of people have reported that meditating with a mantra brings a deeper, more profound experience.

Sometimes mantras are personalized according to criteria that a teacher has been trained in (such as a person's age, date of birth, or various psychological predispositions), but there are also mantras for general use. If you want to try mantra meditation, follow the same technique as for the breath meditation just given. As you breathe in and out, silently use the mantra *So Hum*. The usual method is to use *So* as you inhale and *Hum* as you exhale.

Think each syllable slowly and quietly as you breathe. Don't force the thought, and if you get distracted, easily return to the mantra. Some teaching makes the point that mantra meditation shouldn't be tied to any rhythm, even the natural rhythm of breathing. An alternative technique is proposed where you sit quietly and think *So Hum*, then let go of the mantra and think it again only as it arises in the mind. You gently remind yourself to say it regularly, not simply ignoring the mantra. It's a matter of easily giving it preference over other thoughts. Don't set up a regular rhythm, however, and never try to drum the mantra into your head.

Once you've meditated for a set period, it's important to sit still—or better yet lie down—and relax for a moment before returning to activity. Since mantra meditation takes many people so deep, it's jarring to jump up immediately without a period of letting your mind rise back up to the surface of everyday thoughts.

Pranayama: Because breathing is so intimately connected to every activity in the body, you might consider some ancient techniques from the Yoga tradition that center on the breath. Although these can be quite intricate and time consuming when someone sets out to control or direct their breathing, there are also easy forms of

Pranayama, as these techniques are called. The one we recommend is for refining our breathing and adding to the relaxation and calming effect of your meditation.

Sitting upright, you will be gently breathing out of your left nostril and right nostril alternately. The rhythm is to inhale on the right side, then exhale on the left before switching to inhale on the left and exhale on the right. A few minutes' practice makes this quite easy, actually.

First hold your right hand up with your thumb against your right nostril and two middle fingers against the left nostril.

Gently close the left nostril and inhale through the right. Now exhale through the left nostril by moving your fingers away and gently closing the right nostril with your thumb.

Don't move your hand yet, and inhale through the left nostril. Then close that nostril with your fingers and move away your thumb to open the right nostril—exhale.

It sounds tricky when written out this way, but you are essentially alternating the side you breathe on. You may find it easier to get the knack if you begin the first couple of tries by exhaling and inhaling on the right, then switch your hand position and exhale and inhale on the left.

In any case, be easy with your *Pranayama*, doing it for five minutes before starting to meditate. Most people have a dominant nostril that changes throughout the day. Sometimes you are breathing mostly on the right or the left, probably because one nostril is more open than the other. *Pranayama* is supposed to even out and refine the breath. This can feel odd at first, so if you find yourself growing short of breath or wanting to gasp, stop the practice, sit quietly, and resume normal breathing. Never force your breathing using the technique. Each exhale and inhale should be completely natural. Don't try to instill a regular rhythm or to make your breaths deeper or shallower. It takes more discipline to adopt *Pranayama* than to adopt simple meditation, but those who master it report deeper experiences in their meditation.

THE SCIENCE BEHIND THE CHANGES

The genome and epigenetics are beginning to reveal more about how meditation works. In 2014 we tested the effects of intensive meditation by assessing the activity of genes spanning the entire human genome. The study was conducted at a retreat at the Chopra Center located in Carlsbad, California, just outside San Diego.

Sixty-four healthy women from the community were invited to stay at the La Costa Resort for one week—the Chopra Center has its facilities there—and were then randomly assigned either to a meditation retreat or to a relaxation retreat only, excluding learning to meditate. Serving as controls for the study, the relaxation retreat group would basically spend the time just being on vacation. During the week, blood samples were collected from both groups and measured for aging-related biomarkers.

In addition, any changes in psychological and spiritual well-being were also assessed, not only over the week but continuing up to ten months afterward. By day five, both groups actually underwent significant improvements in their mental health and beneficial changes in their gene activities, including lower activity of genes involved in defensive stress and immune responses (you'll recall that inflammation is a defensive response of the immune system). In the control group, these beneficial changes could be attributed to something termed the "vacation effect," in which stress levels are minimized and the genes that usually deal with stress and injury can "take a rest." The body acts as if all is well and can turn all those stress response genes down.

But other changes occurred in the meditation group that did not happen in the controls. For example, there was two- to threefold suppression of a gene activity associated with viral infection and wound healing. There were also beneficial changes in the genes associated with risk for Alzheimer's disease. These changes suggest that it would be more difficult for the meditators to experience a viral infection while at the same time their systems were less concerned about the need to heal wounds or tend to injury.

Perhaps the most astonishing result specifically found in the meditators was a dramatic increase in the anti-aging activity of telomerase. The importance of this change is explained in the newest edition of Deepak's book on the mind-body connection, *Quantum Healing*. In 2008, the heart disease pioneer Dr. Dean Ornish, working in collaboration with Nobel laureate Elizabeth Blackburn, made a breakthrough by showing that lifestyle changes improve gene expression. One of the most exciting changes had to do with the production of the enzyme telomerase (see page 60 for our initial discussion of telomerase). To recap briefly, each strand of DNA is capped at the end, like a period ending a sentence, by a structure known as a telomere. With age, it appears that telomeres weaken, causing the genetic sequence to fray at the ends.

It is thought, with considerable supporting research, that increased telomerase, the enzyme that builds telomeres, might significantly retard aging. The Ornish-Blackburn study discovered that telomerase did in fact increase in subjects following the positive lifestyle program Ornish recommends.

The Chopra Center study amplified these findings by looking specifically at the mental and spiritual component of a changed lifestyle. The Ornish program has several components, including exercise, diet, and stress management. Under the calm and introspective conditions experienced by newly instructed meditators, telomerase began increasing the longevity of chromosomes and the cells that enclose them.

As a baseline, reduced stress during a vacation induces beneficial patterns of health. For those participants who were able to carry out deep and meaningful meditation, however, there were more benefits beyond the vacation effect, including anti-aging, a reduced propensity for viral infections, and the suppression of genes mobilized for injury and wound repair. It's just as important to note that the effects happened quickly, within a few days. This accords with other findings about how rapidly the epigenome can change.

The bottom line: you can't be on a permanent vacation all year, but you can meditate to achieve the same and more results.

The next frontier. To follow up on such an intriguing study, we next created a research project exploring the possibility of inducing even deeper changes. The power of choice, we believe, has infinite potential. We call this project the Self-Directed Biological Transformation Initiative (SBTI). We've gathered together a consortium of top-tier scientists and clinicians from seven leading research institutions: Harvard University, Massachusetts General Hospital, Scripps Clinic, University of California San Diego, University of California Berkeley, Mount Sinai's Icahn School of Medicine, and Duke University. A particular focus is on the health benefits of traditional Ayurvedic practices. Over at least two millennia, Ayurveda has stressed the primary importance of balancing body, mind, and environment to maximize the body's rejuvenating powers. The SBTI employs state-of-the-art scientific methods to test for benefits in well-being from a multifaceted Ayurvedic approach that includes diet, yoga, meditation, and massage. Instead of studying one possible result, we are taking a "whole systems" approach.

Technology has now made this possible. Our controlled trial design uses wearables—mobile health sensors—and calls upon a host of specialized areas of expertise that are expanding explosively today: genomics, cellular and molecular biology, metabolomics, lipidomics, microbiomics, telomerase assays, inflammatory biomarkers, and Alzheimer's biomarkers. We don't need to go into detail about these technologies, each of which involves extensive specialized knowledge. (Also included are personal evaluations of psychological outcomes at the Chopra Center.)

Technicalities aside, it's enough to say that to our knowledge, this is the first clinical study employing a whole-systems approach to lifestyle, and Ayurveda in particular. While traditional medical research is attempting to develop and validate new drugs targeted at specific diseases, we believe that in a parallel effort, it's only prudent to pursue the lifestyle track, for all the reasons we've been developing in this book. To be fully real, radical well-being must step up and deliver valid data, as the SBTI is currently doing.

Brain changes: If you step back a little, what we're discovering is quite amazing—literally the ability of the mind to transform the body, and to do it quickly, with minimal struggle. The mind can even lead to the generation of new brain cells. Beginning in the 1970s, studies had shown that something is happening in the brain during meditation. This parallels the subjective experience of feeling calmer and more relaxed. But in the last decade, the research has begun to show that meditation can also produce long-term structural changes in the brain, especially in regions associated with memory. There is an increase in a person's sense of self and empathy toward others, along with a reduction in stress levels. Increased brain activity starts to appear in subjects who practice mindfulness meditation for only eight weeks. A team led by Harvard-affiliated researchers at Massachusetts General Hospital reported these results in the first study to document meditation-produced changes over time in the brain's gray matter.

What makes this finding so important is that it links how people feel when they meditate with their physiology—the kind of proof that neuroscience demands. The old view was that meditators reported all kinds of mental and psychological benefits when in fact all they were doing in meditation was entering a state of deep relaxation. In the Harvard study, magnetic resonance imaging (MRI) scans were taken of the brains of sixteen participants two weeks prior to the study and directly afterward. MRI images of the participants were also taken after the study was completed. It was already known that during meditation there is an increase in alpha waves in the brain. Alpha waves are associated with deep relaxation. These MRIs showed something more permanent: denser gray matter (i.e., more nerve cells and connections) in specific regions like the hippocampus, which is crucial for learning and memory, as well as in other areas associated with self-awareness, compassion, and reflection.

Another study compared long-term meditators with a control group and found that the meditators had larger gray matter volumes than nonmeditators in areas of the higher brain (cortex) that are associated with emotional regulation and response control. A famous study of Tibetan Buddhist monks showed activity in the area of the brain associated with compassion.

Loss of gray matter (brain cells) and their connections is a common part of aging. Now it appears that this loss isn't inescapable. Some older people appear to be genetically protected from the deterioration of memory and brain cells, but in general only 10 percent of people who believe they have superior memories actually do, according to the standards set for a study of such "super agers." Still, there is much to learn from these people. Finding out what makes them so unusual is a promising line of research, with the main focus being on their brains as compared with those of younger controls and "normal" older people.

MAKING THE SCIENCE WORK

The science is undeniable, but it takes more than science to motivate people, so we return to the core issue of compliance. We believe that success builds on success. You should look for positive changes in your outer life as well as inside. The science tells us that feelings are a reliable indicator that brain changes are actually occurring. The positive input from feeling more successful adds something new to the feedback loop between mind and body.

As yet the link to outer success, which is often reported by meditators, awaits scientific study. You will be striking out on your own. The point is to see if your outer life is beginning to show improvements that only meditation can explain. No one can really judge this besides you yourself. You may even harbor a not-so-secret belief that meditation makes someone weaker, less competitive, and less motivated. Quite the opposite is true.

Here's a checklist of the changes we have in mind. Within a week or two of beginning to meditate, check off any of the following results you are beginning to notice.

WHAT MEDITATION IS DOING FOR MY SUCCESS

__ I'm making better decisions.

__ I feel calmer, less anxious about making a decision.

__ My work is going more easily.

__ I'm in my comfort zone more.

__ I have a strong sense of self.

__ My focus and concentration are improving.

__ My mind has fewer distracting thoughts.

__ I'm not so dependent on outside approval.

__ I'm coming up with better ideas.

__ I have more energy at work.

__ I'm enthusiastic about what I do.

__ I'm more optimistic.

__ I bounce back better from negative events.

__ I'm getting better at reading a situation.

__ Working with others is getting smoother.

__ I'm having more insights.

__ Problems are less discouraging, more like opportunities.

__ I'm coping with stress better.

__ I'm handling difficult people better.

__ I feel more fit.

__ I feel better put together in general.

__ My mood has generally improved.

Studies like the ones conducted by Ornish-Blackburn and the Chopra Center confirm that there's a biological basis for these benefits. They are based on making one of the harder choices: meditating for 20 minutes twice a day. But even if you decide to make an easier choice, such as taking 5 to 10 minutes out of your lunchtime to meditate, you will start to get the benefits of relaxing and rebalancing your system.

One can also rely on the testimony of thousands of meditators over the years. It's a major shift from the Western model of hard work and struggle to succeed. We understand that, but in our view, you owe it to yourself to take advantage of such an important breakthrough.

SLEEP

Still a Mystery, but Totally Necessary

Nothing has changed in decades over the standard recommendation to get a good night's sleep. Medical science still hasn't determined exactly what sleep does, but waiting for the mystery to be solved is secondary. Primary is the fact that going without sleep throws your entire system out of balance. Something seemingly far removed from sleep, such as obesity, is actually linked quite closely. It's now known that the two hormones that regulate appetite, ghrelin and leptin, are thrown out of balance by lack of sleep. When the brain isn't receiving normal signals about hunger, you wind up overeating. Just as crucial, your brain won't know when you've had enough.

In our parents' generation, getting the recommended 8 hours of sleep every night was easier. Americans now get an average of 6.8 hours of sleep, edging under the minimum of 7 hours that's considered healthy. Older adults sleep less, but it's not because they need less. Current findings indicate that a tiny clump of brain cells in the hypothalamus acts as a "sleep switch," and these cells decrease as we age. Previously the cause of insomnia in the elderly was unknown. Now it seems that brain changes are involved, which helps to explain why seventy-year-olds on average sleep an hour less than twenty-year-olds.

Our focus, then, is on insomnia rather than sleep itself. For most people, a diagnosable sleep disorder isn't the problem. In the tradition of Ayurveda, insomnia is rooted in an imbalance of Vata, one of the three *doshas,* or basic physiological forces. Vata, which is linked to biological motion, causes all manner of restless, irregular behavior. When it is out of balance, people find it hard to keep to a routine in diet, digestion, sleep, and work. Mood swings and anxiety are related to Vata. Without asking anyone to adopt an Ayurvedic perspective, we think it's helpful to see that Vata links mind and body in a very realistic way. Appetite, mood, and energy levels are all thrown out of balance when sleep—a natural remedy for Vata imbalance—is deprived.

Here's a chart that shows you how sleep and Vata can go out of balance together.

The Vata-Sleep Connection

Both are thrown off by the following:

Anxiety, depression
Overexertion
Staying up late
Cold temperature
Irregular eating, poor nutrition
Emotional upset
Physical aches and pains
Excitement, agitation
Stress
Worry
Grief
Harsh surroundings
Excess noise

Taking advantage of the Vata-sleep connection, you should first recommit to getting a good night's sleep. Letting a full 8 hours turn into 5 or 6 is a slippery slope. If you have a problem with insomnia, either finding it hard to fall asleep or waking up during the night, don't turn to pills—sleep aids of every sort are not the equivalent of establishing a natural sleep rhythm.

Instead, our menu of choices is about setting body and mind into the right framework for the brain's natural sleep switch to be activated.

Reading the menu: As in every section on lifestyle, the menu of choices is divided into three parts, according to level of difficulty and proven effectiveness.

Part 1: Easy choices
Part 2: Harder choices
Part 3: Experimental choices

Please consult page 120 in the diet section if you need a refresher on what the three levels of choice are about. Ordinarily we say that you should make one change per week total, not one from each lifestyle section. But in the case of sleep, many of the changes are so simple that it's fine to choose several and let them overlap. Still remember, though, that whatever choices you make are meant to be permanent.

Sleep: The Menu of Choices

Circle two to five changes that would be easy to make in your current sleep routine. The harder choices should follow after you have adopted the easy choices.

PART 1: EASY CHOICES

- Make your bedroom as dark as possible. Blackout shades are best. If total darkness is impossible, wear a sleep mask.

- Make your bedroom as quiet as possible. If you can't achieve perfect silence, wear earplugs. These are also advisable if early-morning noises wake you up.
- Make sure your bedroom is comfortably warm and draft free.
- Take a warm bath before bedtime.
- Drink a glass of warm almond milk before bedtime. (It's rich in calcium and promotes melatonin, a hormone that helps to regulate the sleep/wake cycle.)
- Meditate for 10 minutes sitting upright in bed, then slide down into your sleeping position.
- Avoid reading or watching TV half an hour before bedtime.
- Take a relaxing walk before you go to sleep.
- Take an aspirin an hour before bedtime to settle minor aches and pains.
- No caffeinated coffee or tea three hours before bedtime.
- Use the evening hours after work as a time to relax.
- Meditate in the evening after you come home from work.
- Find ways to unwind from stress—see our section on stress (page 145).

PART 2: HARDER CHOICES

- Be regular in your sleep routine, going to bed and getting up at the same time every day.
- Remove the TV from your bedroom. Keep the bedroom a place for sleeping.
- Attend to signs of anxiety, worry, and depression.
- Don't take work home with you.
- Get a massage before bedtime from your spouse or partner.
- No alcohol in the evening.
- Buy a more comfortable mattress.

PART 3: EXPERIMENTAL CHOICES

- Experiment with herbs and herbal teas traditionally associated with good sleep: chamomile, valerian, hops, passionflower, lavender, kava kava (note that these are not scientifically proven remedies).
- Cognitive therapy (see page 191)
- Get tested at a sleep disorder clinic.
- Sesame-oil massage (see page 191)
- Ayurvedic herbal remedies for Vata imbalance (various over-the-counter formulations are available by mail or at health food stores)

EXPLAINING THE CHOICES

The Vata connection links most of the choices to conventional insomnia advice in Western medicine. Only a few things need further explanation. To begin with, the overlooked things that keep many people awake are too much light in the bedroom, too much noise, and minor aches and pains that escape notice until you start to go to sleep. If you have the kind of sleeplessness that's typified by waking up in the middle of the night or too early in the morning, attend to these three factors as first-line remedies.

The tendency to lose sleep as we age has a Vata link, since according to Ayurveda, this *dosha* increases with age. It's prudent not to take sleep for granted even if you have always enjoyed good, sound sleep. Adopt our recommendations early and you will prevent future problems. Lack of sleep has been associated with triggering Alzheimer's—see page 288 for a fascinating discussion of the Alzheimer's-sleep connection, which Rudy played a major part in solving. Lack of sleep is also associated with high blood pressure, which tends to increase by the decade as we age.

Massage is very relaxing, of course, and if you have a very cooperative spouse or partner, they might be coaxed into massaging your neck and shoulders for a few moments at bedtime. Ayurveda advises *Abhyanga*, a specific daily massage with sesame oil, to settle Vata. It's a simple if slightly messy procedure. Warm a few tablespoons of pure sesame oil (found in health food stores, but not the darker kind used in Asian cooking). Sitting down with a large towel on the floor to catch drips, lightly massage the oil into arms, legs, neck, and torso.

It's not necessary to apply more than the faintest film of oil, and the best time is in the morning after you bathe or shower. Abhyanga is considered the sovereign remedy for Vata, and in addition is a good preventive for catching Vata-related diseases like cold and flu, but it requires a good deal of commitment as something you intend to keep up permanently.

Cognitive therapy has sometimes been effective for those who have long-standing insomnia. In such cases there's almost always a psychological price to pay. Lying awake in bed is unpleasant and discouraging. Insomniacs grow increasingly frustrated. They hate the lack of energy and blurred thinking that lack of sleep brings with it. Cognitive therapy seeks to reverse the negative thinking that has built up as a result of so many negative associations with sleeplessness. Check and see if you find yourself fitting the following mental patterns and behavior.

Fearing the coming night, certain that you won't fall asleep again
Disliking your bed and bedroom
Worrying that you aren't sleeping at all
Tossing and turning in frustration
Obsessing about not getting to sleep
Feeling victimized
Blaming every woe on your insomnia
Staying up too late because you know you're not going to get to sleep anyway
Getting up in the middle of the night to read or watch TV

These ingrained habits of mind and behavior make insomnia worse, so it's worth experimenting with a few cognitive steps you can take on your own, short of seeking help from a therapist or a sleep-disorder clinic. First comes some positive thinking that the latest science upholds.

- Most insomnia is temporary and stress related. It goes away when daily life gets less stressful.
- Insomniacs do in fact get to sleep at some time during the night, even when they think they haven't.
- REM (rapid-eye-movement, or deep-dreaming) sleep is a state that can be reached fairly quickly, even in a short afternoon nap.
- Contrary to previous beliefs, you can catch up on sleep deficit by sleeping in longer on the weekends.
- The brain can remain alert on short sleep for a few hours. With as little as 6 hours of sleep, you can be normally alert and functioning for a while before impairment starts to set in.

Focus on these positive thoughts in order to remove some worries about your insomnia. Become realistic about the actual problems it's causing; don't pile on new or imaginary ones. Make it your goal to stop fixating on lack of sleep, turning your energies to solving the problem. Second, to overcome a sense of being victimized, write down a list of things you are going to do to solve the problem, and then follow through with them. Third, don't let your spouse or partner contribute to the problem by keeping the light on after you want to go to sleep, snoring, or moving around too much in a crowded bed. If you can't sleep separately for whatever reason, enlist your partner in helping you solve this problem.

If you take up insomnia as a challenge rather than an affliction, your frame of mind will change. The solutions we've suggested are many, and countless people in your position have learned how to get a good night's sleep. There's no reason why you can't, too.

THE SCIENCE BEHIND THE CHANGES

Science's inability to explain either the mechanisms or the purpose of sleep has been reduced to a medical school punch line: "the only well-established function of sleep is to cure sleeplessness." To date, the research on sleep has been focused on the brain more than on the genome. We know that brain activity changes during sleep, and some basic discoveries, such as the need for REM sleep, emerged decades ago. It's also becoming clear that when normal sleep deteriorates, it is a subtle sign that other things are going on. For example, some people who suffer from severe depression report that the first sign of an oncoming attack is that they no longer sleep well. By attending immediately to their irregular sleep, they can sometimes prevent the attack from arriving.

It has also become clear that sleep rhythms differ from person to person. In sleep research terminology there are "larks" (early risers) and "owls" (late risers) whose sleep habits are set for life. How such habits get set isn't known, and this may be a fruitful area for epigenetics to explore, since it's through epigenetic marks that genetic predisposition intersects with experience. Disrupting someone's natural sleep rhythm is known to have widespread implications for the body. Workers on the night shift, for example, never fully adapt to their unnatural schedule of waking and sleeping. About 8.6 million Americans work the night shift or rotate shifts, and they are at higher risk for cardiovascular disease, diabetes, and obesity. Since the same conditions are associated with inflammation, there could be a strong link there.

Society may also be paying a price by setting the school day too early. Teachers complain that middle school students are in such a drowsy state early in the morning that they are essentially sleeping through first and second periods. Adolescents need more sleep than adults, between 8 and 10 hours, but one study found that only 15 percent of teenagers get 8½ hours or more of sleep per night. Forty percent get 6 hours or less. The typical adolescent pattern of keeping irregular hours and staying up late leads to problems that are easily preventable. The ideal time for an adolescent to go to sleep is 11 p.m. This implies that school should start later. A national debate has started among educators around the subject. At least one school district experimented with starting the school day an hour later and found that test scores rose significantly for middle school students.

Science would benefit by knowing why we actually need to sleep. Does the brain have to rest for a while? Is it resetting itself, or perhaps going into a mode in which it heals potential damage or grows new cells? Evidence points in many directions. Freud's theory that dreams are disguised messages about the state of a person's unconscious doesn't seem to be valid, according to modern psychiatric understanding (there are holdouts, of course). Current belief is that dreams and the images they produce are essentially random. But this, too, is open to conjecture. Neuroscience can hardly improve on Shakespeare's observation after the guilty Macbeth cannot fall asleep. "Innocent sleep. Sleep that soothes away all our worries. Sleep that puts each day to rest. Sleep that relieves the weary laborer and heals hurt minds. Sleep, the main course in life's feast, and the most nourishing."

Any full understanding of sleep must have roots in how we evolved, however. That much is certain; therefore, genes are crucial in some way as yet unknown. Deepak coauthored an article on sleep with an academic expert, Dr. Murali Doraiswamy, professor of psychiatry at Duke University. Because the genetic links between human and animal sleep are fascinating, we thought we'd give you some of the basic insights, even though they don't apply in any practical way to how you are sleeping.

Their article notes that babies spend most of their days sleeping, but why? Why do creative solutions sometimes arrive in our sleep or soon after waking? ("A problem difficult at night is often solved in the morning after the committee of sleep has worked on it."—John Steinbeck) Do plants go through rest cycles that are the equivalent of sleep?

Such puzzles have been made more topical by a recent study in mice that showed that one of the roles of sleep may be to clear out the accumulated garbage from the brain. If this is the only explanation, however, then why do we need to spend one-third of our day unconscious—couldn't evolution have developed a system to clear out trash while we are awake (much like urination or defecation)?

Let's take a look at some facts that may help us grapple with sleep's mystery and its insights. Sleep is a state in which the organism's consciousness is reduced or absent, and it loses the ability to use all nonessential muscles (in deep sleep you are essentially paralyzed and cannot move your limbs). From birth to old age there are dramatic changes in the amount of time humans spend in various sleep stages as well as in overall sleep. Babies sleep for 15 or more hours, which then steadily decreases to 10 to 11 hours for children and adolescents, 8 hours for adults, and 6 hours for the elderly (even though they need the same 8 hours as when they were younger).

The amount of time spent in REM versus non-REM sleep also decreases through life. Premature babies spend almost all their sleep (some 75 percent) in REM sleep, whereas full-term babies typically spend about 8 hours nightly in REM, which drops to about 1 to 2 hours nightly in adults. During REM sleep the brain shows high activity (gamma waves) and high blood flow, sometimes even more than while awake, and scientists believe this is when the brain rehearses and consolidates actions and memories. One can only wonder what a newborn baby, who spends 8 hours in REM sleep, is dreaming about since it has had so little waking experience.

Most animal species studied appear to sleep. Many primates, such as monkeys, sleep as much as we do, about 10 hours. Dolphins and some other marine creatures can sleep with half their brain awake (unihemispherical sleep) to protect themselves from predators—total sleep of both sides of the brain may lead to drowning. There is still debate about whether or not migrating birds may be able to sleep even while flying (with one eye open, much the way humans can take a catnap while standing up). For whatever reason, at least in captivity, carnivores (such as lions) need more sleep than herbivores (such as elephants and cows)—we don't know if the same applies also to meat-eating versus vegan humans!

All of this interesting stuff illustrates how sleep is programmed into our genes and behaviors. But sleep would seem to be a poor survival trait as far as evolution goes. Because sleep put our ancestors (and other living creatures) at risk from predators, the benefits must outweigh the risks—that's all that scientists can manage to agree on. Unlike humans, some animals (e.g., newborn dolphins) can survive sleep deprivation for a couple of weeks without apparent harm. However, in most species, after extended sleep deprivation, their body temperature and metabolism become unstable and they die. The longest period a human has survived sleep deprivation is believed to be about two weeks, but many physical and mental deficits occur long before that; driving ability is significantly impaired after one night's bad sleep.

Finally, sleep is related to mood—strangely enough, sleep deprivation can make people happy and sometimes manic. Decades ago doctors took advantage of this fact in trying to treat depression (a misguided strategy, now that we recognize the link between depression and bad sleep). Numerous creative breakthroughs have been attributed to dreams, such as the tune for the Beatles song "Yesterday" (Paul McCartney), the structure of carbon and benzene (August Kekulé), and the sewing machine (Elias Howe). Indeed, the discovery of acetylcholine, a chemical that regulates many aspects of dream sleep, reportedly came to Otto Loewi in dreams on two consecutive nights in 1921. On the first night, he woke up and scribbled down some notes in his diary that, alas, he couldn't read in the morning. On the second night, he was lucky enough to write them more legibly. Loewi's subsequent experiment based on his dreams won him a Nobel Prize. Even Rudy had dreams that helped him find one of the Alzheimer's genes based on historical photographs adorning the walls of Massachusetts General Hospital near his lab.

Common experience tells us to agree with Shakespeare's simple conclusion that sleep "knits up the raveled sleeve of care." Without a fuller understanding of consciousness itself, however, all arguments are adrift in the same darkness we inhabit every time we fall asleep.

MAKING THE SCIENCE REAL

When it comes to applying the science of sleep, at this point you may say, "What science?" But there are certainly enough data on sleep deprivation to underscore the need for a good night's sleep at all ages, with deleterious consequences if you don't get the proper sleep. Don't fool yourself into believing that you have trained yourself to do well on less than 7 hours of sleep a night—only a fraction of the adult population falls into this category.

And the genetic connection? We know that the daily, or circadian, rhythm of sleep is maintained by "clock" genes that operate by sophisticated feedback loops. An entire network of these clock genes displays rhythmic activity, although once again, how this activity occurs is unknown. Certain variants in clock genes have been associated with whether you are a morning or an evening person. Attempts to link sleep disturbances with neuropsychiatric disorders have led so far to the identification of gene mutations in clock genes that are associated with rare sleep disorders.

Epigenetics has also been shown to regulate our circadian rhythms and may in fact be closely linked to sleep disorders. Since disruptions of sleep rhythms have been linked to numerous disorders, such as Alzheimer's, diabetes, obesity, heart disease, cancers, and autoimmune diseases, we must further explore the epigenetic link to sleep regulation.

Progress is being made. A specific clock gene called CLK serves as master regulator of our sleep cycle by epigenetically switching other circadian rhythm (sleep cycle) genes on and off. The fact is that hundreds of genes follow a twenty-four-hour cycle of variable activity, and many of these genes affect your sleep cycle and thus your health. Since epigenetics has already been shown to modify the activities of these sleep cycle genes, it follows that a variety of lifestyle changes that affect our epigenetics most likely influence our sleep cycle.

It will be very important to understand which lifestyle activities, experiences, and exposures allow us to sleep regularly or cause sleep deprivation. At the very least, stress must be involved in sleep deprivation. We've previously discussed how stress is a major contributor to epigenetic changes leading to disease. But here we run into a chicken-and-egg question, because sleep deprivation leads to stress and vice versa. More epigenetic findings are in order.

Our recommendations about curing insomnia are also useful if you already enjoy normal sleep, because they can improve its quality. The brain's sleep switch is geared to two activities that are the opposite of each other: arousal and relaxation. Arousal keeps us awake, and it wakes us up if we're asleep. If a loud bang in the middle of the night wakes you up, that's an example of arousal; so is bright light striking your eyes or a dripping faucet.

These external triggers can be managed with a little effort, but there's the subtle problem of internal arousal, which is more difficult to manage. When worry keeps you awake at night, that's an example of internal arousal—the brain refuses to relax, let go, and stop thinking. Some internal triggers are physical, such as when pain wakes you up in the middle of the night, or the need to empty your bladder. We think that the Vata connection is useful here, because Ayurveda takes for granted that body and mind work together, which is certainly true when it comes to sleep.

In Western terms, arousal triggers send too many signals to the brain's feedback loop. Worry, anxiety, and depression are self-perpetuating. Unless a way is found to break their repetitiveness, the same thoughts return obsessively, which interrupts the signal that the brain should be heeding, the signal to go to sleep. The Ayurvedic advice to avoid overstimulating your mind before bedtime is sound advice for our physiology. Stimulus leads to arousal. It's easy enough to make your evenings more relaxing under normal circumstances, but anxiety and depression pose their own special difficulties. This is especially true when someone has become so habituated to worry or negative thinking in general that the brain's sleep switch has become sidelined, as it were.

The opposite of arousal is relaxation, an activity that modern people reserve for the fringe of their day. They relax when there's time left over from work instead of making it a primary activity. What's needed is a new model of how a high-functioning brain should operate. What can be done to counter the tendency to seek more and more stimulation while drastically depriving ourselves of relaxation?

The most credible version of the fully integrated brain is the one laid out by a Harvard-trained psychiatrist and neuroscientist, Dr. Daniel J. Siegel, now at the UCLA School of Medicine, who has made a career of examining the neurobiology of human moods and mental states. In our book *Super Brain* we enthusiastically endorsed Siegel's basic insight that the brain needs a whole "menu" of activities during the day. Please see that discussion for a full discussion of this topic. Here we want to highlight three choices on the menu that too many people are deprived of: "in" time, downtime, and playtime.

We alluded to spending "in time" every day in the section on meditation. As the name says, this is time you spend going inside and experiencing the mind at its calmest and most peaceful, but also its deepest. "Downtime" is spent not thinking about work and duty, simply "vegging out" for a while. Lying on your back in the grass staring at the clouds is the ideal kind of downtime. "Playtime" needs no explanation, but how many of us take a few moments to be playful, to laugh and have fun, every day? Siegel's research indicates that adding these overlooked brain activities has enormous curative effects when a patient seeks psychotherapy. Their brains aren't fully functioning because of the lack of certain activities that are totally necessary for a complete and fulfilling life, including normal moods and emotions.

It's only a matter of time before overstimulation is connected with epigenetic changes and inflammation. Rather than waiting for science to catch up, look at your daily life. If you are exhausted by the end of the day, if you are run ragged with no time to relax, if you don't laugh or feel the enjoyment of simply being here, these are signals that you should pay attention to. Sleep holds its mysteries, but the benefits of relaxation and the risks of overstimulation are quite clear. As you tip the balance away from arousal, your brain will return to a natural state of balance, and the results can't help but improve your sleep.

EMOTIONS

How to Find Deeper Fulfillment

Emotions are a vast subject, but there's one statement that holds true for everyone. The most desirable emotional state is happiness. Even though happiness is a mental state, the body is deeply affected by our moods. Chemical messages tell every cell how you feel. In its own way a cell can be happy or sad, agitated or content, joyous or despairing. The super genome amply confirms this fact. If your stomach has ever tightened from fear, the "gut brain" is eavesdropping on your emotion, and when depression afflicts several generations in a family, epigenetic marks may be playing a key role. Most polls find that around 80 percent of people describe themselves as happy, and yet other research indicates that at best around 30 percent of people are actually thriving, while rates of depression, anxiety, and stress continue to rise.

It is highly unlikely that a "happiness gene" will ever be discovered. The new genetics tells us that in complex diseases like cancer hundreds of separate genetic mutations are likely involved. Emotions are much more complex than any disease. But we don't need to discover the happiness gene. Instead, we should give as much positive input to the super genome as possible, trusting it to produce positive output. Science may take decades to correlate the complex gene activity that produces happiness; in the meantime, the super genome connects all the input that life brings us.

Let's contrast the kind of input that promotes beneficial gene activity with the kind that creates damage. Both lists contain items you are quite familiar with by now, but it's good to see everything gathered together.

POSITIVE INPUT TO THE SUPER GENOME

12 things that reinforce happiness

Meditation
Love and affection
Satisfying work
Creative outlets
Hobbies
Success
Being appreciated
Being of service
Healthy food, water, and air
Setting long-range goals
Physical fitness
Regular routine free of stress

It's hard to imagine that someone whose life contains these things on a daily basis wouldn't be happy. By the same token, the things that the super genome reads as negative must be avoided.

NEGATIVE INPUT TO THE SUPER GENOME

14 things that damage happiness

Stress
Toxic relationships

Boring, unsatisfying work
Being ignored and taken for granted
Constant distractions during the day
Sedentary habits
Negative beliefs, pessimism
Alcohol, tobacco, and drugs
Eating when you're already full
Processed foods and fast food
Physical illness, especially if painful
Anxiety and worry
Depression
Unhappy friends

The two sides of human experience constantly vie for our attention, and it must be admitted that for most people, the scars of negative experience are hard to heal. Adding positive input certainly helps—if you were unloved as a child, being loved as an adult makes a huge difference. But happiness will never be bioengineered. Until we reach Part III, on consciousness and the genome, the mystery of emotions will remain a mystery. The lifestyle choices we offer are all worthwhile. Make no mistake about that. But the trail of clues leads farther.

Reading the menu: As in every section on lifestyle, the menu of choices is divided into three parts, according to level of difficulty and proven effectiveness.

Part 1: Easy choices
Part 2: Harder choices
Part 3: Experimental choices

Please consult page 120 in the diet section if you need a refresher on what the three levels of choice are about. You should make one change per week total, not one from each lifestyle section. Remember that whatever choices you make are meant to be permanent.

Emotions: The Menu of Choices

Circle two to five changes that would be easy to make in your current lifestyle in regard to emotions. The harder choices should follow after you have adopted the easy choices.

PART 1: EASY CHOICES

- Write down five specific things that make you happy. On a daily basis, consciously do one of them.
- Express gratitude for one thing a day.
- Express appreciation for one person every day.
- Spend more time with people who are happy and less time with people who aren't.
- Set a "good news only" policy at mealtimes.
- As you go to sleep at night, take a moment to mentally review the good things that happened that day.
- Fix a weekly date night with your spouse or partner.
- Do one thing a week that brings someone else a moment of happiness.
- Make leisure time creative; go beyond watching TV and surfing the Internet.

PART 2: HARDER CHOICES

- Set a worthy long-range goal and pursue it. Best is a lifelong goal (see page 207).
- Find something to be passionate about.
- Cut back on exposure to bad news in the media—make do with one news program or reading one story online.
- Use the positive and negative input charts (pages 202–203) every day.

- Whenever a situation makes you unhappy, walk away as soon as you practically can.
- Don't unload your negativity on others; seek sympathy and compassion instead.
- Do one thing a day that brings someone else a moment of happiness.
- Learn to deal with negativity after you calm down, not in the heated or anxious moment.

PART 3: EXPERIMENTAL CHOICES

- Write down your personal vision of a higher life.
- Find one self-defeating habit and write down a plan to overcome it.
- Explore the time in your past when you were happiest and learn from it.
- Set out to improve your emotional intelligence (see page 214).

EXPLAINING THE CHOICES

Well-being depends upon happiness, yet most people don't really make this connection. Instead they allow their emotional state to drift. Deepak was recently consulted by a woman in her late fifties who insisted that she led a lifestyle carefully groomed to eliminate the wrong food. She exercised regularly, and she was a high achiever who owned her own business and loved the work she did. So why was she afflicted with aches and pains, along with chronic insomnia, exhaustion, and a faintly depressed mood all the time?

It took half an hour to itemize all the particulars of her lifestyle, and then Deepak asked a simple question, which revolved around the woman's insomnia. It was obvious that getting only six hours of sleep at night was causing almost all of her problems.

"What have you done to make your sleep better?" he asked.

"Nothing, really," she replied. The woman had already revealed that her husband snored, the dog greeted dawn by jumping on the bed, and the slightest noise outside woke her up. Deepak pointed out some simple remedies, but she was barely listening.

"Wait a second," Deepak said. "Do you think it's important to take care of yourself?"

She hung her head. "I know I'm not good about that."

"But you're meticulous about so many things, like your diet."

She looked even guiltier. "I do that for my family. Without me, they'd eat anything."

Now the picture was clear. She was a person who burdened herself with everyone's well-being but her own. Self-sacrifice was woven into her personal idea of happiness. The problem was that she had carried it too far. She forgot herself along the way and would carry any amount of stress because this fit her conception of being a good wife and mother.

The solution in the short run was to get her to do something about her insomnia. The solution in the long term was harder, though. She had to retrain herself to believe that her own happiness mattered. She had allowed her emotional state to drift, and therefore she wasn't really connected to any real state of well-being. Her good marriage and high achievement were being undermined, and so were the positive lifestyle practices she attended to so carefully.

All of us put up with significant pain in our lives without trying to make changes. That's why our easy choices are about turning your attention to what makes you happy, and actually thinking about specifics every day. You need to experience what it's like to appreciate another person, for example. Appreciation, like love, isn't theoretical. The actual feeling must register in the brain, and once it does, the mind-body feedback loop has something real to process.

When you take a moment at night to revisit the good things that happened to you during the day, you reinforce every positive experience. By consciously reminding yourself, you retrain your brain. A kind of filtering process is taking place. You select only the things you wish to reinforce, filtering out the mundane, irrelevant, and negative things. Once this becomes a habit, you will begin to experience a real shift in your personal reality. It will amaze you how much has been overlooked or taken for granted. Life isn't good by itself; you must respond to it as good.

In the harder choices, we ask you to go deeper into what makes you happy on the inside. We all receive a barrage of media trying to convince us that consumerism leads to happiness, but very little messaging that points in the right direction, toward happiness as an inner state. This is another reason for making conscious choices—no one will do it for you. Only you can wean yourself off the twenty-four-hour news cycle, which inundates us with negativity. Only you can find something to be passionate about.

Unconsciously, you've cluttered your mind with years and years of experiences that deposit memories of tragedy, disaster, disappointment, and frustration. In the Vedic tradition (India's ancient wisdom tradition) these memories reside in *Chit Akasha* (literally, "mind space"), and it's in your *Chit* (consciousness) where you build a self. There is no separate compartment for thoughts, memories, and experiences that are objective, impersonal, and therefore selfless. Like a sand dune gathering a billion grains of sand, the winds of your life have deposited particles of experience in *Chit Akasha,* where they've become part of you. A sand dune has no choice but to be a passive collector of any debris that blows its way, but you can choose not to expose yourself to experiences that constitute negative input—refer back to the chart on pages 202–203.

Worthy goals: On a day-to-day basis, the most valuable choice on the menu is probably the recommendation to consult the lists of positive and negative inputs. Reminding yourself to maximize the positive and minimize the negative goes a long way. Yet we place a great deal of importance on being happy for life, and that depends more than anything on setting a worthy goal that can be fulfilled over a long time span. Momentary pleasure doesn't have nearly the impact of a goal that you spend years to achieve, with every step adding more meaning and purpose to your existence.

What is your worthy goal going to be? This is a unique and important decision. For some people, raising a child toward a fulfilled adulthood is deeply satisfying, or developing a passion for charitable works. There are goals as lofty as reaching higher states of consciousness or as practical as building a family-owned business. You don't have to decide once and for all. Your goal can and should evolve. The key to finding a goal that will sustain you for a long time is to be self-aware. Lasting happiness is tied to knowing who you are and what you are here to do.

No one is capable of being all things. In India, the one pursuit that will allow you to thrive in your life is known as Dharma. Dharma comes from a root word that means "to uphold." If you are in your Dharma, the universe will uphold you, so it is believed. But each of us must test this theory out for ourselves. Modern people are fortunate to have the freedom to find their own Dharma; in the Indian tradition, the choice was basically limited to the work your father and mother did. But the principle remains the same: seek inner fulfillment and the path will be smoothed. The opposite is to put so little value on our happiness that you settle for lack of fulfillment. No one who settles can expect life to bring them much support; dissatisfaction only magnetizes more dissatisfaction to itself.

Dharma can be broken down into smaller compartments. Let's do that now. Think about your worthy goal. For Deepak, it is *service*—call this an umbrella term, a single word or phrase that embraces many smaller, specific things, such as giving freely of your time, thinking of what others need, sympathizing with someone else's problems, acting unselfishly, and so on. Rudy's umbrella term is *positive transformation*—with the goal of leaving this planet a healthier and happier place than when he entered it. You can pick your own umbrella term. Among the possibilities that may inspire you are the following:

Love and compassion for all
Bringing peace and reducing violence
Improving education to decrease ignorance and lack of knowledge
Pursuing creativity
Protecting the weak and dispossessed
Promoting culture and tradition
Exploration and discovery in an area rich for those things
Being of service without judgment against anyone

Most people can find a worthy goal within these categories. Choose a goal without worrying that it must be permanent. Sit quietly and center yourself. Take a deep breath, exhale. Another deep breath, exhale. Now a third breath, exhale.

In your calm, centered state, think about the goal you want to achieve. Let's say that you want to be of service. Ask yourself the following questions:

Am I already living my goal, even if it only occupies part of my time?
Is this activity really enjoyable to me?
Does it come easily and naturally?
Does it energize me rather than take away my energy?
Does it make me feel more like the person I want to be?
Am I in the right situation to keep pursuing my goal?
Do I have a sense that this activity is allowing me to grow?

These seven questions are critical for finding your avenue of greatest happiness, your Dharma. When you can answer yes to them, you are perfectly in your avenue of success. There are other things to learn and more skills to perfect, but you have done something invaluable: You have made success a living reality, an activity that will allow you to thrive today and tomorrow, not some distant day in the future.

THE SCIENCE BEHIND THE CHANGES

The new genetics comes at an opportune time, because from a psychological viewpoint, happiness is at a crossroads. As a science, psychology and psychiatry have spent most of their history trying to heal mental disturbances; in other words, curing unhappiness. By now, though, most people have heard about the field of positive psychology, a name that sounds quite optimistic. In reality, some of the most publicized findings in positive psychology are pessimistic. They include the following:

- People are bad predictors of what will make them really happy. After getting more money, a bigger house, a new spouse, or a better job, they aren't nearly as happy as they wanted to be.
- Happiness tends to be accidental and short term. An experience falls out of the sky that makes us happy for a while, only to vanish or feel stale and boring.
- Permanent happiness is a fantasy. If you are very fortunate and almost everything goes your way in life, you may achieve a kind of steady-state contentment, but it will fall short of being happy all the time.
- There is a set point for happiness inside each of us that we can change only temporarily. After any strong experience, whether positive or negative, we return within six months to our set point, and attempts to change it will most likely prove futile.

These are discouraging conclusions, but fortunately they are all provisional. Human nature is too complex to be reduced to a few hard-and-fast principles. The saving grace of positive psychology is that it sets happiness as a normal goal that we can train ourselves to reach. Despite your emotional set point, which returns you to your normal state of happiness or unhappiness, it's estimated that 40 percent of a person's happiness depends on the choices he or she makes.

We believe this number is too low, because it doesn't take into account the new understanding of epigenetics and how experience gets marked down in our genes, not to mention how the epigenomes of our parents and grandparents affect us. Even less understood is how the microbiome is related to happiness, but at least we know that the "gut brain" is constantly sending an enormous amount of input to the brain itself.

We've described how stress can lead to epigenetic modifications that are detrimental. Fear can also cause epigenetic modifications to the genome. An intense fear reaction, sometimes paralyzing, takes place when someone has a phobia. Whatever induces the state of panic—spiders, heights, open spaces, the number *thirteen*—isn't germane. It's the brain's response that creates the phobia. Recent studies suggest that the phobic response can be addressed at the level of gene activities. Researchers in Australia have identified which mammalian genes are modified when someone feels overwhelmed with fear. As with complex diseases like cancer, the picture is complex. In rats, nearly three dozen separate genes undergo epigenetic modifications in response to anxiety-provoking conditions. As a result of these studies and others like them, we now have a good idea of the genes that control the fear response in humans. Can these same genes be therapeutically targeted to alleviate phobias? The future will tell.

On the other side of the coin, positive emotions, especially love, can also change gene activity. In the animal kingdom, many species mate for life, including wolves, French angelfish, bald eagles, and even parasitic gut worms. One such creature is the tiny prairie vole. But when researchers investigated it closely, they were surprised to find that when prairie voles mate, gene activities change to trigger monogamous behavior.

In species that favor monogamous behavior, including our own, couples are likely to build homes together and share parental responsibilities. A specific neurochemical, oxytocin (popularly called the "love hormone"), has been associated with bringing on monogamy. As it turns out, when prairie voles mate, they turn up the activity of the gene that makes a protein in the brain that inserts itself into the surface of the nerve cell and serves a receptor for oxytocin. Such receptors bind with neurochemicals so that they can elicit their effect on the cell. In other words, even when oxytocin isn't increased or less is available, it is more likely to have an effect on nerve cell circuits now that more receptors exist to bind with it.

The act of prairie vole mating accomplishes such changes by altering gene activity. Further studies have shown that epigenetics are at play in male prairie vole behavior. In these studies, genes for the oxytocin receptor, as well as the genes for the receptor for another neurochemical called vasopressin, were turned on to make more receptors. Vasopressin is known to make male voles spend more time with their mates and more aggressively protect them from other males. However, when the same genes were artificially turned up with drugs, the voles didn't undergo these genetic changes or become monogamous. The desired results could be obtained artificially only if the males and females were allowed to spend six hours together in the same cage before the drug was given. The implications of this study are profound—instead of seeing brain chemistry as a one-way street, with a hormone like oxytocin dictating behavior, it turns out that brain chemistry needs the right kind of behavior in place as well.

Animals bond while human beings love. Different as those behaviors are emotionally, is the epigenome playing a critical role in both? In prairie voles, the oxytocin receptor gene was turned up by removing methyl marks from the gene. This leads to the desire for monogamy, and endocrinologists link it to feelings of love between a human mother and her newborn baby. In stark contrast, oxytocin receptor genes with too many methyl marks, which turn them off, is associated in human beings with autism. (In addition, specific mutations in the oxytocin receptor gene have also been associated with autism.) All in all, epigenetics has a profound effect on the oxytocin receptor, and if the prairie vole offers clues to human behavior, oxytocin helps us to become monogamous.

Clearly lifetime coupling can't be genetically induced by the act of making love in humans. But is there a bond at the genetic level? Perhaps it requires first getting to know each other, as with the voles. Many neuroscientists already accept that oxytocin and vasopressin are necessary for people to bond with a mate and feel love. Certain neurochemicals stimulate areas in the brain used for getting pleasure as a reward, which creates a desire for more reward. This mechanism is involved in the action of cocaine, which stimulates dopamine receptors, potentially leading to a cocaine addiction.

There are people who describe themselves as being addicted to love. Besides the direct chemical effect of oxytocin, as pleasurable feelings are recalled and desired through the oxytocin reward center, love can indeed become an addiction.

But pleasure, in all its forms, can't be equated with happiness. If you put food in front of a hungry animal, it will eat, and brain scans will show that the pleasure center in the animal's brain has been activated. In a human being, emotional responses complicate the issue. When cranky two-year-olds refuse to eat, they can be very obstinate about it. In restaurants some people are extremely picky about what they choose from the menu, and depending on our mood, we can refuse food out of grief, distraction, anger, worry, and frustration. Human reactions depend on chemical messages, but there are so many of these that no one has found a simple chemical formula for happiness. We are the only creatures who respond to stimulus X with any response you can imagine. Brain chemicals serve the mind, not the other way around.

MAKING THE SCIENCE REAL

Happiness is a very new branch of genetic research, and there are ethical reasons why human subjects can't be subjected to extreme emotional states. Our menu of choices is based on the best science available. Bringing positive input into our life is a major step and, fortunately, your mood is very likely to be improved when you address all the other lifestyle choices. Indeed, if a lifestyle change doesn't make you feel happier, it won't stick around for long.

But we are brought back to the mystery of emotions and the fact that unlike animals, merely feeling pleasure isn't enough to make us happy. What is enough? Twenty years ago there was a fad for a newly discovered kind of intelligence, measured not by IQ (Intelligence Quotient) but by EQ, for Emotional Quotient. The key finding was that a person's IQ is separate from their ability to handle emotions intelligently. Although some best sellers emerged that urged the importance of emotional intelligence, there's no accepted standard for this. The most commonly accepted test for emotional intelligence, given to 111 business leaders, didn't correlate at all with how their employees saw them. Thus the link between EQ and superior leadership ability—or superiority in any field—is up in the air.

We think a stronger argument can be made for emotional intelligence and happiness. Consider the following desirable emotional traits:

Seven Habits of High-EQ People

1. They have good impulse control.

2. They are comfortable with delayed gratification.

3. They can see how someone else feels.
4. They are open to their own emotions.
5. They know how emotions work and where each one leads to.
6. They successfully feel their way through life instead of thinking their way though.
7. They meet their needs by linking with someone who can actually fulfill them.

All of these traits would allow you to process your experience in a happier way, and it's the processing that counts. You can process any event—a new baby, winning the lottery, moving to a new house—as a source of happiness or unhappiness. Human emotions don't follow rules, which is why we are both creative and unpredictable. But within each person, there must be a way to comfortably relate to how you feel. To us, that's the great benefit of emotional intelligence.

Let's see how each desirable trait might apply in your own life.

1. Impulse control

Consumerism would collapse overnight if people didn't act on impulse. Unthinking choices lead us to stop off at McDonald's instead of eating a home-cooked meal that we know in advance will be more satisfying and healthy. On impulse we eat, drink, and spend too much. Like anything else you train your brain to do repeatedly, impulsiveness becomes a habit, and once entrenched, it's very hard to supplant.

The root of impulsive behavior is lack of control. Most impulsive lapses are harmless, because we all like to lose control once in a while. But beyond that, losing control means that your impulses control you. Past lessons are never really learned if you can't apply them the next time you have an irresistible urge. People with high EQs are the opposite. They learn from the past, and the primary thing they learn is that impulsive behavior is mostly self-defeating.

This is a lesson they actually feel. Their memories don't go blank when it comes to how bad a hangover feels, or being overstuffed after a meal, or finding out that a time-share was a worthless purchase. In fact, having an emotional memory, which most people avoid, is something they are proud of. The memory bank of impulsive people is filled with terrible decisions they prefer to forget; the memory bank of people with a high EQ is filled with good choices that reinforce the next good choice.

What to do: Delay your impulsive action by waiting five minutes. If you still feel impulsive, take a piece of paper and write down the pros and cons of your impulse. Be sure to include how it felt the morning after you indulged in your last impulsive behavior.

2. Delayed gratification

Older people are often heard decrying that the young want instant gratification, but the key is to know which pleasures should be delayed and which can be enjoyed right now. It's gratifying to move out of your parents' house, get your own place, and support yourself. Going to law school or medical school delays this gratification for years and burdens you with considerable debt on top of everything. Society makes it easier to make such a choice because it holds out the promise of prestige and higher income after you graduate.

As we alluded to before, it's largely the small choices where people find it hard to deny instant gratification. That's why we find ourselves

Eating between meals
Overindulging in alcoholic beverages
Snacking while we watch TV
Sitting at home instead of getting some exercise
Stopping off for fast food
Loading up on sugar
Spending hours online instead of relating to real people

Blurting out things we later regret
Going on bad dates instead of waiting for someone better to come along

As with impulse control, which is closely related, people with a high EQ don't seize on instant gratification. They aren't motivated by an intellectual notion that this is good for them, or not entirely. They feel better when they postpone their pleasure under the right circumstances. They are flexible enough to lay down no hard-and-fast rules. Being flexible comes with a high EQ. When faced with a momentary temptation, they don't say, "I'll give in just this once. What can it hurt?" which is naked rationalization. Instead they say to themselves, "Is this really the best I can do? Let's wait and see."

What to do: Take a good look at your life and ask if you have been making problems by seeking instant gratification. Do you waste money on pointless purchases? Is your closet crammed with too many clothes? Is impulsive spending lowering your bank account? Is your freezer full of food you never get to?

If you see a problem, address it one activity at a time. When you are being tempted by a new pair of shoes, for example, or an extravagant purchase like a full-scale home gym that will soon be gathering dust, write down something further off that would bring you even more pleasure. Instead of the shoes, you can save for a vacation. Instead of the expensive gym, you can learn to play tennis and use public courts. Until the delayed gratification is brought to mind, it can't compete with instant gratification.

3. The ability to empathize

It comes naturally to see how someone else feels. We've all had this ability since infancy, when our feelings depended heavily—sometimes solely—on how our mothers felt. Families are the schoolhouse for everyone's emotional education, and of course some children are much more fortunate than others. They don't learn bad habits that must later be unlearned. If you can't easily see how someone else feels and know why, somewhere along the way you blocked an ability you were born with. Either you had a teacher, like a closed-off father, who motivated you in the wrong direction, or you made up your own mind that emotions aren't a positive aspect of life. In any case, you no longer empathize.

People with a high EQ do. It gives good doctors a naturally comforting bedside manner. It makes people come around to a sales pitch, because they feel that their needs are being understood. At some level none of us can be fooled by insincerity and hypocrisy; we have extremely sensitive emotional gauges. With a high EQ, you find it easy to read someone else, looking beyond their words to how they actually feel.

What to do: To empathize with someone else, you have to want to. With people we love, this is easy—when our children hurt, we hurt. Extending this response to someone we like is also fairly easy. Knowing that you have the seed of empathy inside, you can choose to let it flourish. Listen to a stranger or a coworker as if they were friends. Notice how well they respond, then check your own response. If extending sympathy doesn't feel good, there is resistance somewhere inside you. Perhaps you feel that other people's problems impose a burden of responsibility on you. You might feel compelled to help or to worry about them.

Emotional intelligence is about coming to terms with these obstacles and turning them into virtues. It's good to help others, but you don't have to help everybody. It's empathic to listen to another person's story, but not over and over. Once you start to make these distinctions, you'll find that empathy is a wonderful gift, not something to shun or be anxious about. There's a happy medium between the extremes of being too soft hearted and too hard hearted. Set out to find the balance that works for you.

4. Emotional self-acceptance

Being completely open to our own emotions is rare. Inside everyone is a desire to be seen in the best light, so we avoid exposing negative emotions, even to ourselves. But there is another force inside that counters this desire, a voice that reminds us of our guilt, shame, and bad deeds. Constantly telling yourself how good you are is just as far from reality as constantly telling yourself how bad you are. People with a high EQ have confronted the best and worst about themselves. As a result, they find self-acceptance at a much deeper level than most people.

Because we are so defensive about the parts of ourselves that provoke guilt and shame, finding self-acceptance isn't easy or instantaneous. "Love yourself" is the goal, not the first step. Even to say "I am worthy of love" can be quite difficult for some people. They don't have a foundation as well-loved children, which is how we adopt our ingrained sense of self. It's helpful to realize two truths. First, having an emotion you don't like isn't the same as acting on that emotion. Nonetheless, guilt and shame don't see any difference. They want to punish you just for having a thought. In reality, thoughts come and go; they are transient visitors, not aspects of your core self.

Second, you are not the same person that you were in the past. Guilt and shame don't believe this—they constantly reinforce the message that you haven't changed and will never change. In reality, you are constantly changing. The real issue is whether you want to reinforce who you are today or who you once were. People with high EQ find vitality in being themselves here and now. They don't haul in withered selves from the past.

What to do: Anytime you have a guilty or shameful thought from the past, stop and say to yourself, "I am not that person anymore." If the feeling returns, say the words again. Sometimes such recurrent thoughts are very stubborn. In that case, as soon as you can find a moment alone, sit with eyes closed, take a few deep breaths, and center yourself. We aren't minimizing that wounds from the past can have a powerful influence over the present. The key is to realize the falseness of applying old hurts to new situations. With this conviction in mind, you can move toward self-acceptance day by day. Entering fully into the present moment is the best way to find self-acceptance, and vice versa. The more you accept yourself, the richer the present moment will become. Make this truth work to your advantage.

5. Emotional consequences

All actions have consequences, including emotions. As far as your brain is concerned, generating the neurochemicals that give you the feelings of anger, joy, fear, confidence, or any other feeling, is an action. Your whole body reacts to these chemical messages; therefore emotions can't be seen as passive. Even a stoic who bottles up every unwanted emotion is doing something active. In this book we have been focusing on system-wide choices that bring benefits to mind and body together, using the super genome as their vehicle.

Once you know that negative emotions are harmful to you, your viewpoint changes. It's no longer a free ride to attack someone else, feel envy, act out of spite, and fantasize about revenge. Each of these emotions rebounds on you, right down to your genes. True well-being isn't possible when negativity is undermining it. People with a high EQ have come to terms with this truth, even if they don't know about epigenetic modifications. Other people have certainly experienced how a parent's anger or worry caused their children to suffer. On that basis alone, you can grasp that emotions always have consequences.

What to do: You can't stop our negative feelings from having an effect, both on you and on your surroundings. When this fact sinks in, taking responsibility for your emotions is the most important step. There is no longer a valid reason to vent your anger at others, to make them afraid of you, to intimidate, bully, or dominate out of selfish motives.

No one is asking you to become saintly. Knowing that emotions have consequences is meant to benefit you. Open your eyes and watch how someone's anger or anxiety changes the atmosphere for the worse. Feel it in yourself. Then ask if this is the effect you want to have. Emotions are alive. You have to negotiate with them, and when an emotion sees a benefit in changing, it will—you will.

6. Feeling your way

Because so many people distrust their emotions and, particularly with men, try to hide them, it's a shock to hear that feeling your way through life works better than thinking your way through life. In fact, this notion is so alien that we feel the need to point out some strong psychological findings as support for it.

First, researchers have discovered that emotions are part of every decision we make. There is no such thing as a purely rational decision. When you try to eliminate feelings from the equation, you are repressing a natural aspect of yourself. Do you spend more when you are in a good mood? You may not think so, but studies prove that good moods loosen up the pocketbook. Will you pay too much to feel more important, to look better in a salesman's eyes? Many people will.

One of the most intriguing findings in this regard focused on an auction where subjects were asked to bid on a twenty-dollar bill. There was some confusion and laughter over this game. It seems obvious that nobody would bid more than twenty dollars for a twenty-dollar bill. But they did. Especially with males, winning the auction and beating out the other guy were more important than rationality, and so the bidding went higher and higher until someone gave up. Of course, the "winner" had made a ridiculous purchase, but emotion was able to override reason.

People with a high EQ don't shirk from the emotional component of decision making. They are in touch with how they feel, and thus they tap into the deeper aspects of intuition and insight. Once you let your emotions comes to light, you don't have to act upon them (which is the chief fear of repressed people who can't bear the thought of letting their emotions get away from them). The next step is to realize that emotions possess intelligence, and beyond lies a deeper trust in intuition. Emotions unlock whole departments of consciousness that most people are unaware of. For every "gut feeling" that turns out to be right, there are countless other signals being sent to us every day that we need to feel, not analyze.

What to do: If you are already used to feeling your way through a situation, everything we've just said seems obvious, but that's not true for someone who distrusts emotion. Learning to be guided at the feeling level means taking one small step at a time. To get started, think about all those times when you pushed your gut feeling aside and went with your head, only to say later, "I knew that would happen. Why didn't I go with my feeling?" This isn't a rhetorical question. The reason you didn't go with your gut feeling is that you haven't trained yourself to.

The next time you're conflicted between all the reasons to do something and the simple fact that your emotions tell you not to, write down what each aspect of yourself is saying. Then act, following your head or your gut. When the situation resolves and you've found what the outcome is, go back and consult what you wrote down. This works best with people, because we all have some interaction—going on a blind date, working for a new boss, talking to a car salesman—in which feelings can't be ignored and could make the difference between success and disappointment. If you note in writing what you felt, it becomes much easier to trust your intuition the next time. Repetition is the key, as well as looking with open eyes at how often your feelings turn out to be right.

7. Meeting your needs

When you have a need, who do you go to? Let's be specific. You struck up the courage to say something difficult, and the person you talked to shot you down. You are hurt and discouraged. The sting of their words rings in your ears. What you need at that moment is solace and sympathy. If you go to a friend who listens politely, murmurs a platitude or two, and quickly changes the subject, you've turned to the wrong person. You wouldn't go to a donkey for milk, so why did you do the equivalent in emotional terms?

The answer is complicated, but it involves emotional intelligence. When they hurt, most people are so desperate to unload their pain that they turn to the next nearby person. If they happen to be married, they will almost certainly turn to their spouse. But someone with a high EQ will know who is a sympathetic listener and who isn't. They will turn to the one and avoid the other.

Consider a deeper need, the need for love. When this need is fulfilled in childhood—a critical part of having emotional intelligence—being loved has come from the appropriate source, one's parents. But parents can be withholding and unloving, which creates emotional confusion. You grow up not knowing who can actually give you the love you need, and then what happens? You experiment rather randomly, going from one person to the other without being able to see who is capable of love. When you find someone who isn't, who has a bit of love to give, but not very much, you are likely to choose that person anyway. A combination of insecurity, neediness, and emotional wounding leads you into relationships that turn out to be frustrating, disappointing, and in the worst cases, toxic.

Finding the right person to fulfill your needs is so basic for people with a high EQ that they are baffled when someone doesn't. But the sad truth is that wounded people mostly seek out other wounded people or even people who are likely to hurt them. They are often made anxious by the behavior of someone who is emotionally healthy, because it threatens the isolated, closed-off emotional existence they are so accustomed to. Yet the effort must be made; otherwise, we stumble through life feeling enormously unfulfilled.

What to do: Most people find themselves somewhere between dating, courtship, marriage, and divorce. The gap between having a need and getting it fulfilled is something they understand. In all relationships, you can't ask someone else for something they don't have to give. We find ourselves doing that anyway, asking for sympathy from someone who is indifferent, for understanding from the self-centered, for love from the emotionally stunted, and worse.

Yet the steps to solving this dilemma aren't as difficult as you might suppose. When you feel a need, turn to the person who you are certain is able to fulfill it. Who is that person? You can only know if you've seen them respond in a similar situation. Don't guess. Don't take a stab in the dark. People who are kind, loving, emotionally generous, and understanding don't hide those traits. They live by them.

You will soon find that most people want to be there for you. Who hasn't found a pleasant stranger on an airplane who winds up listening to our family situation, romance, work, and even our deep secrets? There's an impulse to hold back, naturally, out of fear of rejection. But it's not hard to first detect a sign of openness and then to take it one step at a time. A little openness leads to more, and if you see that the other person has no more to give—no more time, advice, sympathy, or interest—you will take the hint.

The only caveat is this: even someone who has love, sympathy, compassion, and understanding to give also has the right to say no. We realize how hard this is to accept. Rejection is the biggest reason that most people shy away from encounters that contain any emotion. It's easier to share your trouble with a familiar friend or family member who sits there like a blank wall. Blankness is better than no. But needs are meant to be fulfilled, and you must develop the courage to find the right people, even though you risk being sent away.

The chances are that you won't be, however. Not every need is for undying love. The most common need is to be listened to, followed closely by the need for sympathy and the need to be understood. Validation is the common thread. Once you've discovered that you can be validated—and deserve to be—you will be stronger inside. Then asking for love becomes much easier.

Emotions evoke powerful responses, and all the needs we've talked about lead to changes in the body. Science lags behind wisdom in this area. As a species we've had thousands of years to become wiser, an achievement not to be disparaged because everyone plays the fool at times. We look forward to the day when genetics finds the magic combination of genetic modifications that lead to wisdom. For now, the best guide is our emotions, which may keep ahead of science no matter how much genetics tries to catch up.

Part Three

GUIDING YOUR OWN EVOLUTION

THE WISDOM OF THE BODY

The super genome has liberated our thinking about the body, so can it do the same for the mind? Absolutely. No longer is the brain a castle in the air where the mind lives alone. Everything you think and feel is shared by the rest of your body. The brain doesn't say something in English like "I'm bored" or "I'm depressed." Everything is chemical and genetic. The same language is understood by every cell. Whatever happens in the brain is reflected in the exquisitely integrated activities of every cell.

We are in the habit of believing that only the brain has awareness of you and your surroundings. This belief has to change, because it can't be denied that the whole body is intimately interconnected. Not just brain cells, but every cell's knowledge has been honed for hundreds of millions of years. Of course, as soon as you say that a kidney cell is conscious, traditional biologists, who are wedded to the belief that biological interactions can only be random, will cry "foul!" If you go on to say that a gene or a microbe is as conscious as you are, many other scientists will be up in arms.

But to be outraged by such notions isn't good science. One of the most brilliant pioneers in quantum physics, Erwin Schrödinger, said, "Consciousness is a singular that has no plural . . . To divide or multiply consciousness is something meaningless." We are so used to separating mind and body that merging them into one field of consciousness isn't acceptable, but physics has known for over a century that everything in the physical universe emerges from fields, whether it is the electromagnetic field from which light emerges, the gravitational field that keeps your feet on the ground, or the quantum field, the ultimate source of matter and energy.

Imagine right this minute that every cell is as conscious as a person. This would demote the brain from its privileged position. We would have to abandon our belief that thinking is strictly mental, involving a stream of thoughts, images, and sensations inside the brain. But clearly there is a different kind of thinking—nonverbal, without visual images, possessing no voice—that silently upholds every cell. This cellular intelligence has been called the wisdom of the body. To make a leap in your state of well-being, it is necessary to do only three things:

Cooperate with your body's wisdom.
Don't oppose your body's wisdom.
Increase your body's wisdom.

Even a few years ago this kind of language sounded like poetic license. *Wisdom* is a high-flown word we reserve for venerable sages and teachers. In modern life, it's not even a word we tend to use very often. But we aren't dealing in metaphors here. Wisdom is knowledge that comes only with experience, and your cells have plenty of it. Every lifestyle choice we've recommended comes down to one thing: obeying and restoring the wisdom of the body. We've used the vocabulary of genetics so far. Let's see if that vocabulary can be expanded to embrace the body's wisdom as one thing—a field of consciousness—rather than bits and pieces. This will set the stage for the most exciting possibility of all: influencing your own evolution and that of your children, perhaps even your grandchildren.

WISE CELLS, WISE GENES

Cells are faced with many challenges. If you erase all of science's sophisticated knowledge, a cell is like a water balloon that happens to be alive. But it can be endangered exactly like a water balloon. A puncture would let out all the water inside; getting too hot would cause it to burst; too cold and it would develop ice crystals poking through the skin. A water balloon and a cell both have to worry about staying intact in the face of a cruel, ever-changing environment. Over the eons, cells set out to solve this extremely tough challenge.

Their solution is known as homeostasis, the ability to preserve a steady state "in here" no matter what's happening "out there." At first homeostasis was primitive. One-celled organisms evolved to have ion pumps (for chemicals like sodium, calcium, and potassium) on their outer membrane that could keep the right chemical and fluid balance inside them. The next step was to become mobile, so that they could swim after food, escape predators, and head toward a temperature and light level that was best for their survival. The fact that cells aren't simply water balloons but incredibly complex life-forms is the result of solving the whole problem of remaining balanced "in here."

Now jump ahead to the present moment. Your cells still "remember" how the solution works, thanks to DNA. Genetic memory, working over vast periods of time, ensures that no cell, however primitive, turns back into a water balloon. Having learned the trick of cell division, during which each strand of DNA makes a perfect duplicate of itself, life-forms marched forward. Memory was evolution's greatest invention—a totally invisible one—and once it appeared, it had no reason to stop. Cells started remembering more and more things, developing more and more skills, as we do via our brain.

At this moment, with the help of your genes, your cells remember how to keep you alive, an achievement science barely comprehends, because it takes so many dynamic, interlocked, perfectly synchronized events just to keep the chemical balance inside a heart, liver, and brain cell. Although programmed by the same DNA, heart, liver, and brain cells perform dozens of tasks unique to themselves. In the new genetics, we have to think of the body as a community with 100 trillion inhabitants (adding all of our body's cells to the vast, teeming genes in the microbiome), each of whom has its own self-interest. A heart cell has too much to do on its own to step in for a liver cell, yet this game of "Me first" manages to be about sharing and cooperating as well, because if the heart cell gets tired of messages from the liver or brain and hangs up on the conversation, it dies.

Homeostasis, which started out turning a water balloon into a cell, had to become a billion times more complicated as more cells were invited into the community. Yet in essence, DNA kept repeating the same lesson: stay in balance, preserve a steady state "in here." To show you how essential that is, consider prisoners who go on a hunger strike, as happened during "the Troubles" in Northern Ireland when members of the IRA used such strikes as political protest. The body can remain in healthy balance for only three days drawing upon its reserves of blood sugar (glucose) in the bloodstream and liver. Then it begins to take sugar from your fat cells, and after three weeks or so, it turns to the muscles, which begin to waste away. Starvation mode sets in when the muscles are emaciated, and death becomes inevitable starting somewhere around 30 days, assuming that nothing but water is taken during that period. Mahatma Gandhi, who fasted to publicize the campaign for Indian independence, performed his longest fast for 21 days. The ten Irish Republican prisoners who attracted worldwide publicity by hunger striking in 1981 survived between 46 and 73 days. (We aren't taking into account a person who is grossly obese and decides to stop eating; there are hospital records of survival for over a year without food when someone has three hundred to four hundred pounds of fat and protein to draw upon.)

Total fasting brings the progressive breakdown of homeostasis, which very soon disrupts normal functioning everywhere in the body and is eventually fatal. And yet the survival period can be greatly extended simply by adding a small amount of sugar and salt to the water that is being drunk during the fast. Fasters who add a bit of honey to their water have gone up to five months before they stopped. It's not just the calories that prolong life but maintaining the cells' ion (electrolyte) balance, the most basic factor that makes even the most primitive cell a living thing instead of a water balloon. (Note: We are not endorsing juice, honey, or sugar water fasting of any duration. The pros and cons of these regimens must be reserved for another time.)

Notice how systematically the body reacts to total fasting, progressing from one strategy to the next to remain in balance for as long as possible. The point we're making is that the mechanism for the most basic survival has been preserved in your genetic makeup for over a billion years, while simultaneously your super genome keeps up with everything you want to do today. Homeostasis is as complex as you are. This implies a much wider view of the mind-body connection. As you think, feel, dream, imagine, remember, and learn from the past while anticipating and planning for the future, your body must accommodate it all in the present while never sacrificing its self-interest, which is to survive, if not thrive, and remain healthy.

A typical cell stores only enough oxygen and fuel to survive for a few seconds, so the fail-safe protections must come from elsewhere—in a word, cooperation. A cell "knows," chemically speaking, that it will get oxygen and fuel from the bloodstream, so it doesn't have to "think" about those things, devoting its "intelligence" to other processes. (We are using quotation marks here to differentiate a cell's naturally occurring intelligence from common usage, which involves the volitional implementation of knowledge by the brain.)

Unless homeostasis is disrupted and you begin to feel something out of the ordinary (e.g., pain, dullness, fatigue, depression), the fail-safe mechanisms of the body remain out of sight. But we can relate them to our personal experiences, and as we do, the mind-body connection transcends chemicals and biological processes. Your cells are living the same experiences that you are, sharing the same purpose and meaning. As shown in the chart below, the inherent properties of a single cell are astonishing.

A Cell's Wisdom: 9 Essentials for Life

Awareness: Cells are acutely aware of their environment, meaning they are constantly receiving and responding to biochemical cues. A single molecule is enough to make them change course. They adapt from moment to moment according to changing circumstances. Not paying attention isn't an option.

Communication: A cell keeps in touch with other nearby cells and even some far away. Biochemical and electrical messages are exchanged among cells to notify the farthest outposts of any need or intention, however slight. Withdrawing or refusing to communicate is not an option.

Efficiency: Cells function with the least possible expenditure of energy. They must live in the present moment, but they are totally comfortable with that. Excessive consumption of food, air, or water is not an option. As they attempt to do the most with the least energy, they are constantly evolving to become more efficient.

Bonding: Cells making up a tissue or organ are inseparable companions. They share a common identity through their DNA, and even though heart, liver, kidney, and brain cells lead their own lives, they remain tied to their source no matter what they experience. Being an outcast is not an option. However, renegade cells can create a cancerous tumor.

Giving: The chemical exchange in the body is a constant give-and-take. The heart's gift is to pump blood to the other cells; the kidney's gift is to purify the blood for everyone else; the brain's gift is to keep watch over the whole community, and so on. A cell's total commitment to giving makes receiving automatic—it's the other half of a natural cycle. Taking without giving back is not an option

Creativity: As cells become more complex and efficient, they combine with each other in creative ways. A person can digest food never eaten before, think thoughts never thought before, dance steps never seen before. These innovations depend on the cells being adapted to the new. Clinging to old behavior for no viable reason is not an option.

Acceptance: Cells recognize one another as equally important. Every function in the body is interdependent with every other. Nobody gets to be a control freak. Overstepping needs is not an option; otherwise an abnormality like cancer can result.

Being: Cells know how to be. They have found their place in the cosmos, obeying the universal cycle of rest and activity. This cycle expresses itself in many ways, such as fluctuating hormone levels, blood pressures, digestive rhythms, and the need for sleep. The off switch is just as important as the on. In the silence of inactivity, the future of the body is incubating. Being obsessively active and overbearing is not an option.

Immortality: While cells will eventually die, they are immortal in the sense that they use genes as well as epigenetics to pass on their knowledge, experience, and talents in stem

> cells long after they die. They withhold nothing from their offspring. This is a continuity of existence that is also a kind of practical immortality, submitting to death on the physical plane but defeating it through the propagation of DNA. The generation gap is not an option.

When any one of these nine essentials is disrupted, life itself is threatened. There's no more glaring—or frightening—example than cancer. A cancer cell has abandoned the essentials. Its actions make it virtually immortal on its own by endlessly dividing. It crowds out and kills neighboring cells. It has dismissed the regulatory chemical signals from surrounding cells. Nothing matters but its own self-interest; the natural balance of the cellular community has gone tragically awry.

Oncology is actively deciphering the genetic triggers that are involved in cancer. These are incredibly complex and interwoven. The diabolical truth is that a malignant cell can draw upon the same "intelligence" as every other cell, but genetic mutation directs its activity into madness. Like a consummate criminal, it wildly changes disguises to keep out of the clutches of the police, or in this case, the immune system. If cancer wasn't such a dire threat, such ingenuity proves on yet another front that every possibility the human mind can devise has been anticipated by our cells.

In the face of the incredible complexity posed by the super genome, something simple and useful emerges: the nine essentials that cells preserve at all costs are the same essentials that make each of us human. The mind-body connection is so flexible that it can adapt, not just to adversity but also to perversity—the perversity of turning your back on what Nature has designed you to do, which is to remain in balance. When we submit our bodies to toxins, push them to the point of exhaustion, and ignore their signals of distress, we are flouting the wisdom inside every cell.

On the other hand, we can align ourselves with the same wisdom, and when this happens, the mind-body connection reaches its real potential.

How to Live the 9 Essentials

1. Have a higher purpose that goes beyond yourself.
2. Value intimacy and communion—with Nature, other people, the whole of life.
3. Keep yourself open to change. From moment to moment sense everything in your environment.
4. Nurture acceptance for all others as your equals, without judgment or prejudice.
5. Relish your creativity. Seize on the renewed freshness of today, not clinging to the old and outworn.
6. Feel how your being is cradled in the natural rhythms and patterns of the universe. Embrace the reality that you are safe and nurtured.
7. Let the flow of life bring you what you need. The ideal of efficiency is allowing Nature to take care of you. Force, control, and struggle are not your way.
8. Feel a sense of bonding with your source, the immortality of life itself.
9. Be generous. Commit yourself to giving as the reason for all abundance.

These nine things fulfill the necessity of cooperating with your body's wisdom, not opposing it, and doing what you can to improve it. We've crossed over from lifestyle choices into making your life more meaningful, which is the whole point of well-being. You don't simply want to feel better but to lay the foundation for a fulfilled life.

THE MIND FIELD

We strive to support our arguments with solid science, and seeing the body as a field of *intelligence* is no exception. When someone asks, "Where is the mind located?" most people will automatically point to their heads. Why? It may simply be because so many sense organs are located there: eyes, ears, nose, and tongue. With so much information flowing into one part of the body, it could merely be habit that places the mind in our head. Mind and brain have taken up residence together in a box called the skull. Is the brain actually so closed up in its box that it makes sense to speak of it as though it were a machine for making mind, the way a laser printer makes documents? The new genetics makes us ask some culturally radical questions, including the most radical of all: Is a brain even necessary for all forms of "awareness"?

Evolutionarily, nervous systems are not always centralized. Some creatures, like jellyfish, have neuronal nets distributed throughout the body. While humans do possess a central nervous system, we also have other, more distributed nervous systems as well. We have a peripheral nervous system, which includes nerves that gather information for the brain (e.g., the nerves in our sense organs) and nerves that send signals from the brain (e.g., telling our muscles what to do). After it was observed that the gastrointestinal tract can function quite well when severed from the peripheral nervous system, it was concluded that this constitutes a weblike enteric (intestinal) nervous system.

The deciding factor in calling the enteric nervous system a separate nervous system was the specialized ganglion cells that are located between muscle layers in the intestinal wall; these act like a local brain. If you sever the nerves that contact them from the brain, these ganglion cells continue to instruct the intestine to move and absorb and secrete, working quite well and quite autonomously as a self-contained functional unit.

It turns out that the intestinal tract not only takes advice from the rest of the body. It harbors its own reactions. When bad news gives you a sinking feeling in the pit of your stomach, you are experiencing an emotion as surely as you experience it in your head, and it usually precedes any thoughts you might have. Did the enteric nervous system create such a sensation on its own? That's unclear, but it's tempting to think so. Certainly many people trust their gut reactions over the confused and compromised responses that the brain is often saddled with when overthinking sets in.

Findings about brain-like processes outside the skull have become common. The muscles of your face are directly linked to your brain. While we assume that the brain is telling the mouth and lips to smile when we're feeling happy, the reverse is also true. Seeing a smile on someone else's face can make you happy, and children are taught to smile as a way to break out of a sad mood. Whether this works or not varies from person to person, but it could be argued that the face is controlling the brain in those instances.

It may be that other parts of the body override or rebel against the brain. Rudy, who plays basketball twice a week, has experienced a phenomenon known as "alligator arms." When stressed, distracted, or anxious, the muscle memory of the arm and wrist freezes up, and the ball, shot with the brain's best intentions, can miss the basket by five feet.

The conduction system of the heart, which organizes your heartbeat, can be thought of as the heart's brain in the same way the ganglion cells in the gut are the brain of the intestines. The independence of the heart's conduction system is shown when a transplanted heart keeps beating even though the nerves that connected it to the donor's central and peripheral nervous systems have been severed. The interaction between the heart's independent processing and the brain is complex and not fully understood.

The immune system has been labeled a "floating brain." In a very tangible way, thanks to what is called immune surveillance, your immune cells can "decide" whether an invading substance is friend or foe. If they decide wrong, you develop an allergy to harmless things—house dust, pollen, cat dander—that pose no danger and never needed to be repelled. Ask any allergy sufferer whether their allergy affects their thinking. The dullness, lack of energy, and depleted enthusiasm that many allergy sufferers experience leave little doubt about how the immune system is part of a larger bodily intelligence.

These findings are enough to establish that cultural assumptions about mind and brain are full of gaps. The location of the mind is an open question, and any attempt to isolate it physically in the skull runs into valid objections. More and more it looks as if every organ is the locale of its own version of mind. (You might imagine it like the United States, with a centralized federal government, many state governments, and a myriad of local governments working together and influencing one another.)

Thinking is happening, in some guise or other, everywhere in your body all the time. This emerging view has the potential to rock our accepted understanding of mind itself. The brain looks more and more like an outcropping in a landscape that is permeated with varied forms of intelligence. Let's explore the implications of this new model.

In the old model, nerves were like the wiring that brings electricity to every part of a house. But it's not just the "wiring" of nerves that links brain to body. Hormones and neurochemicals produced by all sorts of organs affect the way the brain works and how you experience your mind. Consider the mood changes experienced by many women around their menstrual period and menopause, or by men during a midlife crisis. Other mental events are triggered in similar biological ways. Ever feel sleepy after you've had too much to eat? Ever feel an adrenaline rush after public speaking or feel addled after you're thrown accidentally from your bike? Hormones travel to the brain via the bloodstream, producing profound effects on the nature of "your mind." A panicky thought created by adrenaline, secreted far away from the brain in the adrenal cortex, feels like "your thought"—biology has mysteriously converted itself into mind.

THE BRAIN OUTSIDE THE BRAIN

Looking at the brain itself reveals even greater complexity to the mind-brain relationship. While people generally think of neurons as the particular brain cells that produce the mind (acting together in almost infinitely complex networks), there are other cells in the brain without which the neurons could not do their jobs—the glial cells, for example, which outnumber neurons and perform many essential tasks: conveying nutrients and oxygen to neurons, creating the myelin sheaths around their long trunks (axons) to facilitate speedy signal transmission, stabilizing connections between neurons, and serving as the immune system to protect cells against harmful microbes. In connection with Alzheimer's disease, glial cells clean up debris from aging or injured nerve cells but can also turn against nerve cells and kill them. This "friendly fire" can occur while trying to protect the brain from invaders like bacteria, viruses, and fungi.

The cells that process mental events aren't necessarily only "of the brain." Neurons can also derive from other resident cells in the body, and some neurons and many glial cells arrive in the brain via the circulatory system—they are like nomads who eventually find a place to live permanently. Questions abound about how much this happens and in which different regions of the brain it is taking place. (The production of some brain cells might occur by circulating stem cells that directly become neurons and glial cells, or by fusing with preexisting cells.) All these issues are still being worked out by developmental biologists. It's clear, however, that cells are trafficking between the body and the brain all the time.

So the boundaries between brain and nonbrain in the body are not clear cut. The brain is permeable to the rest of the body. To say the brain *creates* the mind is at best incomplete. It may be more accurate to say the brain provides access to the mind. In a simple analogy, every automobile needs an engine in order to run. But an engine by itself goes nowhere. The functions that make a car a car require every part acting in concert. Likewise, the functions that our dynamic minds carry out are created by the body-brain complex, not by the brain alone. The brain has always been out of the box; it's just been waiting for science to catch up. Mainstream science is reluctant, if not dismissive, when faced with the notion of mind outside the brain. Actually, getting your mind to move outside your head is relatively easy. If you burn your hand on the stove, your attention immediately rushes there. The heartache of unrequited love takes one's attention to the center of the chest. In various spiritual traditions, this kind of "moving mind" becomes a conscious skill. Here's a common introductory example of "mind outside the box" from Zen Buddhist practice. Students who have taken on a disciplined daily Zen meditation—usually counting or following the breath—are then advised to move their minds into the *hara*. The hara is the second chakra, or subtle energy center, located below the navel, just in front of the sacrum. One way to describe this "moving mind" exercise is to imagine that the mind is located in a drop of honey in the center of the skull (where we usually experience our mind anyway) and then to let the drop of honey slowly descend down along the front of the spine until it finally reaches the hara.

Succeeding in this exercise takes time and a great deal of practice. Initially it can feel as if there's only a little movement, because your focus of attention snaps back into your head like a rubber band. And so you begin again, letting the drop of honey slowly descend, bringing your mind with it. Why do this? One reason is that when your mind moves from inside your skull into a position in front of the sacrum, it can bring a jolt of energy, not unlike the way coffee suddenly energizes your mind a few minutes after downing your morning cup. What might otherwise have been sleepy Zen suddenly becomes awake Zen.

More important, practitioners report that there is an exquisite sense of stability in their mind when it's brought to that location: thoughts still come and go, but they take on a sense of waves rising and falling, or of clouds passing overhead, rather than being like a restless monkey bouncing all over the room. A mind running around in the space of uncontrolled thoughts makes us tired, but it also disguises the potential for having a silent, strong, still mind.

LOSING "MY" MIND

Neuroscience is leery of subjective experiences, but the fact is that practitioners of Zen and other Eastern traditions routinely move their mind out of their head. The experience has been replicated for centuries; it isn't accidental, haphazard, or hallucinatory. With enough practice, someone can move their mind into their little toe, shoulder, elbow, perhaps even across the room. The immediate answer of most neuroscientists is that such a subjective sense of "moving mind" either isn't real or can be explained away as a kind of neurological illusion, like the "phantom limbs" reported by patients after a leg or arm has been amputated. The phantom limb seems to occupy the same space as the real limb that was lost and even experiences pain.

The best rejoinder to this claim is that a whole host of subjective experiences in medicine are self-reported and cannot be measured without asking the patient what's going on. Statements like "I feel a pain here," "I'm depressed," "I'm confused," and "I've lost my balance" can sometimes be traced to distorted brain activity on an fMRI (functional magnetic resonance imaging) scan, but only the patient can relate what is actually happening. The brain scan can't tell someone he's in pain when he says he isn't. (When a bacterium avoids a toxin in a petri dish or is attracted to food, can we claim to know that it isn't feeling some primitive form of repulsion or attraction?)

There comes a time in all contemplative traditions when one's sense of mind and of the ordinary self changes fundamentally, lasting for a moment or a lifetime. In Vedic and Buddhist traditions, these experiences are called *Samadhi*, in which a connection is made with pure awareness at the deepest level. In Hebrew mystical practice this might be understood as *D'vekut*, in Christian practice "cleaving to God." The ordinary thinking mind is left behind, and one arrives at consciousness without content.

Samadhi enters the shadow zone in which "my mind" dissolves into mind itself. Here, reality shifts dramatically. Instead of sitting inside the space of a room, the person sits inside mental space (*Chit Akasha* in Sanskrit). The events that take place are not strictly mental, however. On the inner voyage, time, space, matter, and energy emerge from silence much the way physics describes creation bubbling up from "quantum foam." In our view, the inner experience of meditation, yoga, Zen Buddhism, and the like are not inferior to the data collected on subjective states like pain, feeling happy, or falling in love. Brain scans offer a correlate with these experiences, but it takes a person to have them.

It makes people woozy, sometimes even apprehensive, to discover that there is no boundary between "me" and the whole world. What about the skin? It is portrayed in high school biology class as an impermeable barrier protecting you from invaders assaulting the body from "out there." But the metaphor of the skin as living armor isn't viable. Your skin is a community of your human cells and bacterial inhabitants. Pause and move your hand, observing how the wrist and finger joints move under the skin. Why doesn't the skin break down with all this motion, the push and pull of your fingers closing and extending, your arm bending and stretching? Because the bacteria lining the creases in your skin digest the cell membranes of dying skin cells and produce lanolin, which lubricates the skin (as does the collagen connecting skin cells). How long would "you" and your genome last if your skin were cracking, open to infection just from typing on a laptop or waving good-bye to someone? Fortunately, we are living communities thriving in harmonious interaction driven by the super genome.

The only reason we separate "in here" from "out there" may be biological rather than based on reality. Research is starting to account for the swing between the inner and outer world, a swing we all experience every day. Sometimes we direct our attention to objects "out there," sometimes to mental events "in here." One hypothesis now suggests specific neural activity within two complementary signaling networks in the brain—one is active when you are dealing with the world outside the body (called the task-positive network), while the other, the "default network" (or task-negative network), revs up when your focus is inward, as commonly happens in wakeful rest, introspection, or from lack of significant sensory input. Our brains are thought to alternate rapidly between these two networks, but when deep meditation is performed, they both activate together. In meditation, "inside" and "outside" are no longer opposite and contrary but are experienced as a seamless whole. And gene activity is changing throughout this magnificent process.

THE FINAL FRONTIER

One last boundary keeps mind and body apart, a rigid belief in physicality. The entire setup of the brain is physical. Every action a neuron takes is physical, and so are the coded sequences in DNA that create nerve cells. Thanks to the new genetics, this coding has become far more transparent; with stunning advances in technology, we can view the tiniest alterations in gene activity. Nowhere along the line, however, can you see DNA obey the mind. Thoughts are invisible, and science is leery of anything that cannot be visibly detected and measured. The validity of science is all about measurement, even if it takes an instrument as powerful as an electron microscope to extend human sight.

Yet we know that our minds are at work. The new genetics has helped the cause of invisibility, so to speak, by showing that subjective experiences in life can lead to epigenetic modifications that alter gene activity. In a way, the fact that our bodies change according to how we think and feel is so obvious it doesn't need science to prove it. The whole body responds when someone loses a spouse, a best friend, or a job, and in the wake of grief there can be depression, greater susceptibility to disease, and even a risk of premature death. Your super genome directly reacts to these life changes.

All these changes are regulated by genes, and yet the lure of the physical remains strong in mainstream science. A geneticist will first look at the chain of molecular alterations in DNA, finding more and more complex links, before anything as intangible as the emotion of grief is considered. This limitation is the final frontier that must be crossed. How can that be accomplished?

One angle is the concept of the field, which is basic in modern physics. Everything that happens physically at the level of atoms and molecules (which are observable "things") goes back to fluctuations in the field (which is invisible and "no-thing"). You can see a compass needle point north, but you can't see the Earth's electromagnetic field, which is causing the effect. You can see a leaf fall from a tree, but you can't see gravity pulling it to the earth. Is something like this happening when genes become active?

An intriguing experiment by British molecular biologists in 2009 might illuminate this point. For decades we have known that DNA has the property of repairing itself, which it does by recognizing which parts of the double helix are incorrectly coded, broken, or mutated. When a cell divides and a strand of DNA duplicates itself, recognition is also involved in reassembling the new strand as each base pair finds its proper place. In their experiment, the British team placed separate strands of DNA in water and watched them begin to form round clumps (sphericals) of genetic material. A long sequence of 249 chemical bases (called nucleotides) was marked with fluorescent dye to follow how it attached itself to other bits of DNA inside the clump.

The results were astonishing and inexplicable. Exact matching sections of DNA were about twice as likely to join together, recognizing each other even when they were separated in the water by distances that allowed no physical contact. To a cell biologist, this makes no sense, since it takes physical contact or chemical connections for anything to happen inside a cell. But in terms of the field, the mystery has an explanation. Like a compass obeying the lines of magnetic force encompassing the planet, these strands of DNA could be obeying a "biofield" that keeps life intact.

The research team dubbed the behavior of the DNA strands "telepathic" in the absence of any physical connection that drew them together. The biofield, operating through infinitesimally tiny electric charges, might offer an explanation less supernatural. But recognition is a trait we ascribe to the mind. When you wait at the airport for a friend's plane to land, you recognize who she is out of a crowd of strangers, not by going one person at a time, but simply by knowing who you're looking for. In the same vein, but far more mysterious, an Antarctic penguin returning from the sea with food in its crop can recognize which chick belongs to it, heading directly to the chick among thousands of other penguin chicks.

Something about recognition is basic and defies random choice. This is a property of the mind field that all of us depend on—at this moment you recognize words on a page, not collections of alphabet letters that you sort through to obtain what they mean. Apparently DNA can do the same, because the 249 nucleotides didn't match up one by one; the entire sequence found its mirror image, defying randomness.

GETTING IN TOUCH WITH THE FIELD

This telling experiment helps cross the final frontier, but it doesn't get us entirely past physicality. To do that, we must accept that other, as yet indescribable and unmeasurable, factors are acting behind the scenes, organizing bits of matter into living creatures. In mystical traditions around the world, adepts have experienced this invisible agent.

All that's needed is to contact into your natural field of intelligence, present from your brain down to every cell in your body. Fields are infinite, but you don't have to be. A small horseshoe magnet is an outcropping of the Earth's immense magnetic field, and in turn, the Earth's magnetic field is the tiniest speck in the electromagnetic field of the universe. Yet every trait of this infinite field is present in a magnet. In the same way, you are an outcropping of your mind as part of a larger mind field. This gives you an automatic connection to it. When an experience of the mind field is clear, as in deep meditation, perception changes. Some people who have entered this state of consciousness have reported the following experiences:

- They sensed infinity in all directions.
- Time and space stopped being absolutes—they were seen as purely mental creations.
- All separation ceased. Only wholeness was real.
- Every event was connected to every other, like waves rising and falling on an unbounded ocean.
- Life and death no longer represented a beginning or end. They were merged into the continuum of existence.

These realizations are available to everyone; you don't have to attend a mystics school. There is nowhere to go, in fact, in pursuit of the mind field, because we are surrounded by it, down to our genes. It takes a special angle of vision to make the field show itself. In the Vedic tradition, a text called the Shiva Sutras gives 108 ways to see beyond the mask of matter and discover what lies beyond. One such technique is to see what lies beyond the sky. You can't perform such an act, not physically, but that's not the point. In the attempt to see beyond the sky, something else happens: the mind stops. Baffled by the impossibility of the exercise, the normal stream of thought ceases. At that instant, the mind perceives only itself. No object obstructs pure awareness, and, aha! *That's* what lies beyond the sky.

A fish surrounded all of its life by water cannot know what water is actually like. But if it jumps out of the sea, there's a contrast, and then wetness can be experienced as the opposite of dryness. You can't leap out of the mind field, but you can slow down your mind, and then there's a similar contrast: you can experience what stillness, silence, and the cessation of activity feel like.

Even if you don't practice meditation, which is where the great sages, saints, and mystics found their deep contact with the field, you can still get a glimpse. Sit quietly with your eyes closed, doing nothing. Notice the stream of thoughts going through your mind. Each mental event is temporary. It comes, stays for an instant, and then departs. In between each mental event, notice that there's a brief gap. By diving into this gap, you can reach the mind field in its infinite extent. But you don't have to try it this minute.

Having glimpsed the gap between two thoughts, open your eyes. Consider what you've just experienced. Mental events rise—but from where? Mental events fade away—but where to? The mind field. We pay so much attention to our thoughts that we miss this simple point. Each thought is a transient event, while the mind is permanent and unchanging. Did you feel how easy it is to notice this? For a brief moment you've become a *Gyan Yogi*, someone who is united with the mind field. Or to be more precise, someone who *knows* they're united with the mind field, because there's no such thing as losing contact with the field. We just forget about the field, being obsessed over the mind's constant round of thinking, feeling, sensing, and imagining.

We aren't criticizing the activity of the mind. Experiencing the mind field only deepens your appreciation of life. It engenders the wonder that caused the Persian poet Rumi to exclaim, "We come spinning out of nothingness scattering stars like dust," and on another occasion, "Look at these worlds spinning out of nothingness / This is within your power."

Life evolves according to patterns everyone finds beautiful to behold. Evolution gave rise to the human genome and the brain, the most complex structure in the known universe. Can this mystery be solved by looking beyond the mask of matter? The body exhibits almost infinite "intelligence" in every cell. What we refer to as cellular "intelligence" is the cell's natural ability to adapt, respond, and make the right choices at every moment not only for itself, but in service to every other cell, tissue, and organ in the body. *Something* caused this to happen. In pursuit of that *something*, we need to address evolution itself, the force that makes it possible for all of us to be here in the first place.

MAKING EVOLUTION MINDFUL

The super genome has vastly expanded the idea of the responsive and adaptable cell. It opens the door for many other exciting prospects. A responsive and adaptable cell can modify its DNA as its environment poses new challenges and opportunities. It can receive and interpret messages from the brain and respond in kind. The cell therefore adapts to our life experiences, constantly reorganizing and attaining balance to better serve itself and other cells in the body. What we witness is a mind-body partnership. The human mind is conscious. It uses adaptation, feedback loops, creativity, and complexity in astonishing ways—they are the prized possession of our evolutionary place in Nature. Cells mirror the mind, giving physical expression to it.

There's only one problem with this picture, and it's a big one. The theory of evolution doesn't consider that genes mirror consciousness. Introducing a term like "the intelligent gene" would be anathema, even though most geneticists didn't protest "the selfish gene." To be selfish implies making choices to serve only oneself, and it takes consciousness to do that. Our cells make choices all the time. Imagine a steel pellet moving in a circle on a sheet of paper. The pellet seems to be magically moving on its own, until you look under the paper and see that a magnet is actually controlling it. Something similar seems to be happening with the activity of cells in your body.

Let's say you were somehow able to observe heart cells individually, and for no apparent reason they start twitching like mad, only to slow down a minute later. They appear to undertake this action on their own, but if you step back, it turns out that the person who owns the heart ran up a flight of stairs. The heart cell responded to instructions from the brain, and the brain was obeying the mind. That's how the partnership works. What we deem intelligent is the person, not his cells. Even brain cells come second in the partnership, because mind is always first.

Evolutionary theory finds itself in the reverse position, putting matter first. Mind, as far as conventional modern Darwinism is concerned, evolved from basic cellular activity that was mindless. Chemical interactions became more complex, as did the cell's ability to adapt to its surroundings. Single cells started to clump together to form complex organisms. After hundreds of millions of years, the clumps became specialized, and the central player was a clump that evolved into nerve cells, primitive nervous systems, and finally primitive brains. We know all of this because, lucky humans, our clumped nerve cells stand at the peak of brain evolution. The human brain made us conscious, aware, creative, and highly intelligent.

This book has proposed, to the contrary, that cells and genes participate in the same mind field as the brain. This theory is acceptable to anyone who believes, as Darwinians do, that matter comes first. But our view has one big advantage. It opens up a new frontier in the mind-body partnership. Pandas will never stop eating bamboo shoots; tigers will always stalk deer; penguins will always walk across the Antarctic ice fields to lay their eggs—at least for the foreseeable million years. It would take at least that long for a mutated gene to alter such powerful instinctive behavior.

But human beings can change their diet, renounce violence, become vegetarians, and have babies in a warm hospital instead of the Antarctic. We are endlessly adaptable. Therefore we've pushed evolution far beyond physical boundaries. Our skin radiates heat in such quantity that spending a winter's night outdoors would be fatal to a naked human, yet we've sidestepped such a huge disadvantage through clothing, shelter, and fire. We've become evolutionary oddballs, without a doubt. But our next advance may outstrip anything accepted in mainstream Darwinism.

Human beings could be the first creatures in the history of life on Earth to self-direct where their evolution is going. If so, the super genome becomes the key to everyone's future, starting with what each of us is thinking and doing right this minute.

To get there, however, three major changes would need to be established in our understanding of evolution, and each of them would topple a pillar of Darwinian theory.

First, evolution must be driven by more than random chance.

Second, evolution has to drastically speed up, able to bring changes not in hundreds of thousands and millions of years, but in a single generation.

Third, evolution must be self-organizing and thus *mindful*, allowing for the influence of choice making, learning, and experience.

These are serious challenges to the status quo. Ordinarily the argument would take place within the small circle of professional evolutionists. But the goal is so important to everyone's life that we want to bring you into the privileged circle. As much as any famous geneticist deserves to talk about where human evolution is headed, you also deserve to. Let's examine the three changes to Darwinism that need to occur, not because we two authors say so, but because these are the very changes that may lie ahead, thanks to the new genetics.

IS EVOLUTION JUST A LUCKY BREAK?

We mentioned at the outset that the notion that all new mutations occur *only* randomly belong among the discarded myths about genetics. At that point, the sound of many an angry evolutionary biologist hurling heavy objects across the room could be heard in the background, because far and away the phenomenon of solely random mutations has been a primary tenet of Darwinism. To claim otherwise has been a standard line of attack among anti-evolutionists with a religious agenda, and it's hard to remove that stain.

In Darwinian theory, the mutations that drive evolution aren't driven by life experiences. According to Darwin, a giraffe didn't acquire its long neck because it wanted to have one or needed one. The longer neck appeared accidentally one day, and that lucky mutated giraffe then gained a survival advantage that was naturally selected to be passed on to subsequent generations. It's obvious that a longer neck allows giraffes to reach leaves higher up on a tree, but Darwinism doesn't allow for any "why" to intrude. Classical evolutionary theory doesn't allow you to say a long neck appeared "because" the animal needed to eat higher up on the tree; it would say the new mutation was random and persisted "because" it gave the animal this new ability to survive.

Outside the field of evolution, we talk about "why" and "because" all the time. If a basketball player is three inches taller than everyone else on the court and scores more rebounds, it's because he's got the advantage of height. So why can't we say the same about the giraffe? The reason has to do with how mutations are passed along. That first lucky giraffe had to survive or else its new mutation goes nowhere. Next the mutated gene had to appear in the next generation. If it still gave a survival advantage, the gene was now present in more than one animal—this improved its fighting chances.

But the odds were still hugely against it, because to be permanently established, the mutated gene had to find its way into the genome of every giraffe; the short-necked ones had to be so disadvantaged that they disappeared from the gene pool. The process is a numbers game, pure statistics repeated generation after generation. All that matters is the gene and how successfully it gets passed on. Evolutionists may speculate, using common sense, that a longer neck allowed favored giraffes to get at leaves that shorter giraffes couldn't reach, but that's not all there is to the story, scientifically. The hard data pertain to the persistence of a mutation over time.

Thanks to modern gene theory, the statistics of survival have been honed to a fine degree. Facing the iron wall of random mutations is daunting; you will find the entire genetics establishment rejecting your contrary ideas. At least that was true in the past, up to the last decade. Now the iron wall has become something else, a gap.

A gap is friendlier than a wall, because it only needs a bridge, not a wrecking ball. On one side of the gap we have the obvious fact that human beings are intelligent. On the other we have Darwinian theory, which considers *intelligence* a suspicious term. The term was corrupted by the intrusion of Intelligent Design, a movement that attempted to use science to justify the Book of Genesis. That attempt was foiled by massive protest from the scientific community, and we concur. So we don't need to fight that same battle all over again. The rancorous divide between reason and faith needs to be healed, because both deserve their rightful place.

The gap is starting to close as new findings put pressure on conventional evolutionary theory. Random mutations aren't the whole story, as the new genetics is fast proving. (As the great Dutch-Jewish philosopher Spinoza said, "Nothing in Nature is random. A thing appears random only through the incompleteness of our knowledge.") Natural selection isn't the whole story, either. Unlike giraffes, microbes, and fruit flies, human beings don't exist solely in the state of Nature. We exist in a culture that has deep influences on how the super genome works. If a bad mouse mother can pass on her behavior to her offspring, human behavior could be doing the same, but on a much wider scale.

If the gap between standard evolution and the new genetics can be closed, that's tremendous news for you and every other individual. It means that you are actually evolving in real time, and if that's true, huge things follow.

Can evolution remain intact while at the same time giving up on pure randomness as absolute truth? Can mindful evolution move from Darwinian dogma to established fact? It has to, if the super genome is going to fulfill its enormous promise.

THE FALL OF RANDOMNESS

The evidence that gene mutations are not simply random is steadily mounting. In a 2013 study published in the high-impact science journal *Molecular Cell*, researchers from Johns Hopkins University showed that when mutations are deliberately introduced into yeast to impair their growth, new mutations immediately arise to bring growth back. These are called compensatory secondary mutations. They are anything but random. Compensatory mutations can also arise if the solution in which the yeast is raised is depleted of necessary nutrients, creating a more stressful environment. Although yeast is a very basic organism, the lesson here is that when environmental challenges are evident, the genome can quickly adapt and compensate with necessary (nonrandom) mutations for purposes of survival. Epigenetic modifications of gene activity can be utilized for the same purpose.

Another study, concerning *E. coli* bacteria and published in *Nature,* arrived at a similar conclusion. Mutation rates were highly variable along different parts of the bacteria's genome. Researchers detected a lower rate of mutation in genes with high activity. Contrary to the idea that all mutations are random, the mutation rate among genes appears to have been evolutionarily optimized to reduce the occurrence of harmful mutations in certain genes that are most critical for survival. By the same token, increased rates can be found where mutation is most useful—for example, in immune genes that have to constantly rearrange to make new antibodies as protection against invading pathogens. While it's still not exactly clear how mutations are directed to some genes and not to others when the environment is challenging, a leading hypothesis that's being explored is that epigenetics plays a key role.

Obviously Darwin, living in the nineteenth century, could not have known that mutation rates vary widely along different spots in the genome. He did not even know about the genome. It's getting less and less tenable in the twenty-first century for strict Darwinians to abide by the dogma that mutations occur only randomly and are later subjected to natural selection. The actual rate of mutation at any spot in the genome is affected by multiple factors that vary for the purposes of DNA protection or repair, or by epigenetic factors. This isn't a random process.

Is there enough room in the new genetics to say that each person is evolving at this very moment? Not yet. There are more hurdles to cross, beginning with the speed of evolution, a crawl so slow that species often take millions of years to evolve.

There is also fascinating evidence that cancer mutations are not entirely random, as previously thought. Since the scientific details are rather dense, see the Appendixes, page 282, for a technical discussion of this issue.

SPEEDING UP THE CLOCK

In traditional Darwinism, a species must wait around for a gene mutation to occur randomly. If it promotes survival, the mutation establishes a new behavioral or structural feature in the carrier. It can then take millions of years to spread through the population of the species. But with epigenetics, these changes can happen in large swaths of the population in the very next generation.

Establishing exactly how long it takes for evolution to occur is arguable, and the discussion can begin in many places. Let's start with Darwin's "special difficulty," as he called it, a difficulty that would have far-reaching effects. The problem had to do with ants and honeybees. Darwin could not fathom how sterile female ants continue to show up generation after generation in the colony even though they cannot reproduce. He noted how different the sterile females were in terms of behavior and body shape from the fertile females. Even though the sterile females obviously couldn't propagate and therefore had zero chance to breed, how could their genes keep being passed on? Darwin didn't know about genes, but his theory depended on survival, which isn't possible if an entire class of ants is sterile.

Finding the answer was impossible until the advent of epigenetics, long after Darwin passed on. Epigenetics explains how chemical modifications of DNA can permanently alter gene activity, turning it up or down. This process can happen after the moment of birth, sidestepping the baffling issue of passing on new genes—all that's needed is to modify the existing ones. On his own, Darwin got close to the answer. He speculated that it could be found in the caste systems of honeybees.

Depending on the type of food the honeybee larvae eat, they can be candidates for queen or instead end up as sterile workers in the hive. The difference comes down to a special food known as royal jelly, which contains nutrients that foster greater development of the ovaries. It's been shown that the precise mechanism involves epigenetic alterations of select genes. While the queen bee's diet allows her to live for years and lay millions of eggs, the brief life of a worker bee is relegated to keeping house, taking care of the young, and foraging—basically doing whatever needs to be done for the good of the hive.

A similar mechanism functions in an ant colony. Darwin ultimately went on to propose that in the case of ants, natural selection does not apply only to the individual but also to the family and society. He was beginning to see how an entire colony could be viewed as a single evolving "super organism," which is how we see it today.

Diet can further modify gene activity to program certain honeybees to emit pheromones instructing them either to take care of the young or to go out and bring back food. Gene activity can be modified by the action of enzymes known as histone deacetylases (HDACs), which remove chemicals known as acetyl groups from the epigenetically modified genes. It turns out that royal jelly contains HDAC inhibitors that secure a honeybee's position as a possible future queen. Interestingly, while we were writing this book, the FDA approved the drug Farydak, the first epigenetic drug—a HDAC inhibitor for treating recurring forms of a specific cancer, multiple myeloma (MM). Farydak reverses epigenetic changes that occur on certain genes, with the intention of preventing the spread of MM to other parts of the body.

After 150 years, Darwin's "special difficulty" has led to the realization that epigenetics determines not only the fate of bee larvae, but also their later behavior. This genetic detour speeds up evolution for all practical purposes. Just as important, it makes evolution personal. In standard Darwinian theory, evolution is totally impersonal. To take hold, a new gene mutation must be passed on within a large chunk of the population of plants or animals. The flightless wings of a penguin, for example, allowed the whole species to survive through diving in the sea and swimming continuously after fish. But epigenetics changes the life of the individual. In the case of the honeybee, the entire life of a single sterile female is determined by epigenetic modifications. This difference may have explosive implications for human beings. We've been offering the mounting evidence that epigenetic switching is the key factor in lifestyle choices and well-being. But getting evolutionists to consider, much less agree with, this new scheme meets with considerable resistance.

There is presently a heated controversy over whether *Homo sapiens* has genetically advanced over our relatively brief life as a species. After leaving Africa 200,000 years ago, our ancestors populated far-flung locales around the world, and as they did, the facial features, skin, and skeletal structure of each major group became distinctive. An Asian face doesn't resemble a European face in key ways, just as African skin resembles the skin of neither one of those populations.

As the noted biologist and writer H. Allen Orr explains, "Geneticists might find that a variant of a given gene is found in 79 percent of Europeans but in only, say, 58 percent of East Asians. Only rarely do all Europeans carry a genetic variant that does not appear in all East Asians. But across our vast genomes, these statistical differences add up, and geneticists have little difficulty concluding that one person's genome looks European and another person's looks East Asian."

It's been argued that so much is different from genome to genome that the time line must be sped up to account for it. Some evolutionists believe that up to 8 percent of genetic changes occurred through natural selection in just the past 20,000 to 30,000 years, a blink of an eye in evolutionary time when you consider the rise of the horse, for example, from a small ancestor, *Eohippus* (Greek for "dawn horse"), which was only twice the size of a fox terrier and roamed North America between 48 and 56 million years ago.

In the midst of this controversy, where the data tend to be very "soft" and the conclusions speculative, it's not even clear if our genome changed out of advantages in survival (getting more food) or mating. One camp suggests that genetic changes were not entirely due to random mutations and natural selection but were driven by culture. Because human beings live in collective communities, it is plausible, the argument goes, that traits that promoted community skills were favored through breeding and therefore got passed down to the modern day. But exactly how a gene promotes a specific skill is questionable. It's intriguing to follow the struggle that Yale physician and social scientist Nicholas Christakis went through before publicly stating that "culture can change our genes."

That's the title of an online article from 2008 in which Christakis declares, "I have changed my mind about how people come literally to embody the social world around them." As a social scientist, he had seen abundant evidence that people's experiences—of poverty, for example—shaped their memories and psychology. But that was the limit. As a doctor, "I thought that our genes were historically immutable, and that it was not possible to imagine a conversation between culture and genetics. I thought that we as a species evolved over time frames far too long to be influenced by human actions."

REAL-TIME EVOLUTION

Without using epigenetics to describe why he changed his mind, Christakis gives a striking example of how culture talks to genes:

> The best example so far is the evolution of lactose tolerance in adults. The ability of adults to digest lactose (a sugar in milk) confers evolutionary advantages only when a stable supply of milk is available, such as after milk-producing animals (sheep, cattle, goats) have been domesticated. The advantages are several, ranging from a source of valuable calories to a source of necessary hydration during times of water shortage or spoilage. Amazingly, just over the last 3 to 9 thousand years, there have been several adaptive mutations in widely separated populations in Africa and Europe, all conferring the ability to digest lactose. . . . This trait is sufficiently advantageous that those with the trait have notably many more descendants than those without.

Three thousand to nine thousand years is race-car speed across evolutionary epochs, but Christakis can no longer see any reason for doubt. "We are evolving in real time," he writes, "under the pressure of discernible social and historical forces." These words don't seem dramatic until you realize that "social and historical forces" are to some extent under human control. After all, we start wars, wipe out entire populations, enforce starvation, and, on the positive side, bring relief to famines, cure epidemic diseases, and reform poverty.

The clincher for Christakis was a 2007 article by University of Wisconsin anthropologist John Hawks and his colleagues in the prestigious *Proceedings of the National Academy of Sciences* offering evidence that human adaptation has been accelerating over the last 40,000 years. A sped-up rate of "positive selection," the authors say, can be statistically proven by studying genomes around the world, supporting "the extraordinarily rapid recent genetic evolution of our species." A panorama of possibilities suddenly opened up. Genetic variants may have favored some people to survive epidemics like typhoid after the rise of cities and much closer contact with others.

Once Christakis began thinking this way, he realized that culture isn't speaking a soliloquy, and neither are genes—they have always been in a dialogue. "It is hard to know where this would stop. There may be genetic variants that favor survival in cities, that favor saving for retirement, that favor consumption of alcohol, or that favor a preference for complicated social networks. There may be genetic variants (based on altruistic genes that are a part of our hominid heritage) that favor living in a democratic society, others that favor living among computers. . . . Maybe even the more complex world we live in nowadays really is making us smarter."

Real-time evolution is crucial to the super genome. We can be certain that it's happening in the microbiome, because bacteria live very short lives and are prone to rapid mutations. But if radical well-being is to become a reality, real-time evolution must apply to the whole mind-body system. How would that work? Before Darwinism triumphed, there were other evolutionary theories, and one in particular that foresaw that creatures could evolve in a single lifetime.

The French naturalist Jean-Baptiste Lamarck (1744–1829) was a supporter of evolution decades before Darwin. He was a hero on the battlefield against Prussia and a gaunt, determined figure in the laboratory. He eventually died blind, impoverished, and publicly ridiculed; until very recently, his evolutionary ideas remained an object of scorn, in fact, because they ran contrary to Darwin's. Lamarck proposed that species evolve in accord with the behaviors of the parents. For example, he claimed that if you read hundreds and hundreds of books and become learned, you would then have smart children. Obviously this is not the case. But in view of epigenetics, Lamarck's ideas now appear a little less absurd.

He could be considered the father of "soft" inheritance, which lies at the core of epigenetics—traits that get passed on to the next generation if the mother or father has had a strong enough experience to create epigenetic marks (like undergoing a famine or torture camp) or if the pregnant mother smokes or drinks to excess, or is exposed to environmental toxins. With the tremendous advances in genetic analyses of genomes all the way from the human to the viral, we have validated not just Darwin's theories of "hard" inheritance but also some Lamarckian principles as well. Without being exactly right, he is no longer absurd.

A growing body of epigenetic data says that Lamarck was at least on the right track. Soft inheritance is a prime example of sped-up evolution. Yet it still remains to be proven that lifestyle changes in the parents can be passed on to the next generation. Are they strong enough, and do they persist long enough at the epigenetic level? These are open questions at present. Lacking any knowledge of genetics, Darwin could never even attempt to answer these questions. But some combination of soft and hard inheritance one day will.

BRINGING IN THE MIND

We began this chapter by saying that evolutionary theory needed to undergo three changes for the super genome to fulfill its potential. We've covered the first two, removing the barrier for random mutations and speeding up the rate of evolutionary change. What remains is the third and potentially most controversial point, bringing in a role for *mind*. Since the very word is so explosive, we will substitute terms that describe how systems work when they become highly complex and evolved. There is no use butting heads with arch-materialists—many of them consider mind an offshoot of physical activity in the brain, like the heat thrown off by a bonfire.

We wrote an entire book, *Super Brain*, about the relation between mind and brain, strongly supporting the position that mind comes first, brain second. But a book on genetics must stand on its own. There is no controversy, or little enough, that complex systems are self-organizing, using feedback loops as a form of learning. Learning implies evolution, whether we call it mindful learning or the behavior of a complex system. With that settled, let's proceed.

What would *mindful* evolution look like? It would have direction, meaning, and purpose. The beauty of a brilliant bird of paradise in the New Guinea rain forest, the fearful symmetry of a tiger, the quivering gentleness of a deer—all such traits would be intentional. There would be a reason for them to exist beyond survival of the fittest.

As with other aspects of the new genetics, the absurdity of such a notion has gradually been softened. While it's still a huge leap to claim that evolution has a purpose and a goal (technically known as teleology), it's no longer viable to call evolution totally blind. The pivot occurred when the concept of self-organization began to take hold over the past few decades. When you were a teenager, you probably had a typical teenager's bedroom where the lack of organization is total, with clothes strewn everywhere, an unmade bed, and so on.

But as an adult you faced the need to organize your life, since the alternative is chaos. Evolution was faced with the same dilemma, and becoming more organized in order to avoid chaos brought the same solution.

In 1947 a brilliant neuroscientist and psychiatrist, W. Ross Ashby, published a paper titled "Principle of the Self-Organizing System." His definition of "organization" didn't revolve around its usefulness, the way it's useful to run an organized business instead of a disorganized one. Nor did Ashby judge being organized as good versus bad. He claimed instead that organization pertains to certain conditions among the connected parts of an emerging system. This turns out to have tremendous implications for how our genome organizes itself.

In Ashby's view, a self-organizing system is composed of parts that are joined, not separated. Most important, each part must affect the other parts. The way that the parts regulate each other is the key. A stove isn't self-regulating. If you put on the teakettle and walk away, the temperature gets hotter and hotter until the water boils away and the kettle starts to melt and fuse with the burner. But a thermostat is self-regulating. You can set the desired temperature and walk away, knowing that if the room gets too hot, the thermostat will turn off the heat.

You couldn't survive if your body operated like a stovetop. Processes cannot be allowed to run away with themselves. An unchecked fever of even five degrees above normal human body temperature threatens brain damage and eventually death. Growing too cold shuts down the metabolism and leads to hypothermia, which in extreme cases can also prove fatal. The self-regulation of a thermostat exists everywhere in the body, regulating not just temperature but dozens of processes. Because of self-regulation, you don't grow and keep on growing; your heart rate doesn't speed up and keep on accelerating; the fight-or-flight response doesn't make you run away and keep on running.

Every cell in your body developed through orderly, self-regulated steps, reaching amazing complexity in the fetal brain. In the span of nine months, beginning with a single fertilized egg, nerve cells begin to differentiate, at first in isolation but quickly forming a network. By the second trimester, new brain cells are being formed at the fantastic rate of 250,000 per minute, and some estimates raise this to 1 million new cells per minute just before birth. These cells aren't simply globs of life bunched together. Each has a specific task; each relates to other nerve cells around it; the entire brain knows where every one of its 100 billion cells belong.

Connections, networks, and feedback loops are key to all self-organizing systems. Billions of years ago, early bacteria may have started out independently, but as they encountered one another in the soil, they began to interact and form communities—eventually they totally depended on one another to survive and thrive. In our bodies, as we've seen, bacteria network with our own cells. They share much of our DNA and interact to form an immensely complex and sophisticated microbiome. Evolution has made our survival completely dependent upon them. If in the twentieth century we spent most of our time figuring out how to fight microbes, in the twenty-first century we are focusing on how to harmoniously coexist with them. The super genome is the ultimate self-organizing system, because it reflects the entire history of life on Earth.

DNA, it goes without saying, is incredibly orderly, putting billions of base pairs in order. This is more than ordinary chemical bonding, however. Inside a cell, active self-organizing is going on. Specific chromosomes occupy specific positions in the nucleus. Only 3 percent of the genome is actually made of genes, and the gene-poor regions are near the edge of the nucleus, where there is the least ability for epigenetics to modify gene activity. In contrast, the gene-rich areas of the genome are in the center of the nucleus, where regulation of gene activity is most concentrated. Genes that are controlled by the same proteins tend to cluster together in genomic "neighborhoods," making it easier for those proteins to find the genes they regulate all in one place. Everything we see in the genome says it is not laid out randomly, but logically. With that said, it would be a mistake to then go to the other extreme and say it was "designed" this way. The design becomes apparent only after the fact. The journey there was carried out by the principles of self-organization.

Self-organizing systems exist as their own reasons and cause—they constantly re-create themselves with new interactions. This leads to new states of order that are never complete. For example, an atom is actually a sub-microscopic system that obeys rules of orderliness. Electrons are arranged so that an atom of oxygen is different from an atom of iron. But room has been left for change. Because the outer electrons in these atoms can bond, ferrous oxide—common rust—is created. It, too, isn't completely stable, leading to more changes. Rust is more complex than either oxygen or iron, its two components. Thus complexity fuels greater self-organization, and vice versa.

This is the continuing miracle of evolution, that it defies chaos by making ever greater creative leaps. If you pile up sand on a beach, you get a sand dune. It's massive but not complex; nothing holds it together as a system—one hurricane is enough to disintegrate the dune and make it disappear. But as cells accumulate in a fetus, they don't simply pile up like grains of sand. They bond, interact, and organize. So a strong wind doesn't cause the human body to disintegrate.

But this is just the beginning of the story. Complexity and self-organization, proceeding hand in hand, learned how to create life, and life learned how to think. Set aside for the moment that thinking, as most evolutionists see it, emerged only with the human brain. The entire march of events leading up to the brain shows that new states of order are never complete. As the eminent theoretical biologist Stuart Kauffman put it, "Evolution is not just 'chance caught on the wing.' It is not just a tinkering of the ad hoc, of bricolage, of a contraption. It is emergent order honored and honed by selection."

KEEPING IT TOGETHER

The chemical bond that joins oxygen and iron atoms to make rust is physical, but the operation of your genome contains something that goes far beyond the physical. The technical term for this invisible X factor is *self-referral*. It means that a system keeps tabs on itself by constantly sending messages back and forth so that a circle of change is also a circle of stability.

The key to self-referral is the feedback loop. When a gene makes a protein, you can be sure that either directly or indirectly that protein will help regulate the activity of the gene somewhere down the line. Simply put, if A produces B, B must in some way directly or indirectly govern A. Your own choices, physical or mental, come back to govern you. The scale can be very large or very small. If you are single and decide to marry, this decision puts all your past memories in a new light, just as getting sick puts wellness in a new light and growing old puts youth in a new light. Each phase of life moves forward while at the same time gathering the past around it.

Self-referral is also how your genes can respond with just what's needed for your life today while never losing sight of their programming from the past. At the same time, through mutations and epigenetic marks, the present has the capacity to alter these instructions. This is the basis of self-referral at its very roots. Nothing is produced in the universe without coming back to somehow control that which produced it. In spiritual terms, there is the principle of moral balance between good and evil (the law of Karma), stated in Christianity as "As you sow, so shall you reap." In Newtonian physics, it's the third law of motion: For every action, there is an equal and opposite reaction. Opposites should pull a system apart, but they don't, because the invisible element of self-organization keeps them intact.

Feedback mechanisms underlie the links between an organism and its environment. Allow us to explain this a bit technically, because feedback is such a strong element of the argument. We now know that genes are resilient to forces and counterforces. In evolution, new mutations occur when there is stress and challenge in the environment. When challenging conditions arise, the DNA of certain genes becomes exposed so it can be switched on or off by epigenetics, or turned up or down in activity by specific proteins called transcription factors. This first involves changes in the actual folding and topography of the DNA.

As a result, the exposed regions of DNA can be more prone to mutation. So in this model, which is increasingly becoming more accepted, mutations do not occur in random spots in the genome. Changes in the environment lead to changes in how the DNA is folded (not in the actual sequence of base pairs). This determines which gene regions are exposed to possible mutation. In other words, the environment, life exposures, stresses, and outside challenges affect how the DNA is folded in the nucleus, laying certain regions more exposed to mutation than others. In this case, mutations aren't random but arise downstream of environmental conditions. Even though some speculative thinking is involved here, the feedback between genes and outside conditions is key. It enables an organism to adapt to the conditions that Nature brings. So reliable is this mechanism that it has sustained life from the first primordial micro-organisms onward.

As every component of the genome emerged and interacted with other components, they regulated each other to assemble what appears to be a logical design. But in actuality there was no preconceived design, either historically or in the future. Natural processes achieve their results in real time, through self-interaction. Our minds struggle to grasp how this can happen. Leonardo da Vinci marveled, *"Human subtlety will never devise an invention more beautiful, more simple or more direct than does Nature, because in her inventions noth-*

ing is lacking, and nothing is superfluous." In essence, Nature is all about feedback loops. While our genes set the stage, we determine the character we play on that stage and choose the characters with whom we will interact. And, in return, the set on the stage adapts to us. We are modifying our genes with our words, actions, and deeds all the time. This feedback system has been the cornerstone of evolution and always will be.

MYSTERIOUS INHERITANCE

At a certain point, it seems totally arbitrary, conceited, and human-centric for human beings to claim the mind as our private domain. The notion that Nature mindlessly created our own mind doesn't, at bottom, make much sense. The ingrained cleverness of evolution's stratagems is astonishing, even in so-called lower life-forms. For example, gene-based changes in survival can take place by simple thievery. Take the case of the brilliant emerald-green sea slug *Elysia chlorotica*, which looks remarkably like a plant. When it's time to eat, the slug swipes chloroplasts—cellular machines that can perform photosynthesis—from nearby algae to produce food for itself, the way a plant does, making sugar from water, chlorophyll, and sunlight.

This interesting case of chloroplast burglary has been known for decades, but more recently it's been discovered that the crafty sea slug can also steal whole genes from the algae. These allow it to make its own food. Normally the stolen chloroplasts last only so long, but the genes the sea slug steals and binds onto its genome keep them going strong, producing meals far longer. It's astonishing that an animal can feed itself like a plant through cross-species thievery of genes.

Something similar pertains to our species, too. Scientists used to believe that all the cells in our body contain identical genomes. But we are now finding that more than one genome can be found in the nucleus of a single human cell. More specifically, some people have been found with groups of cells that contain multiple gene mutations occurring nowhere else in their body. This can happen when the genomes of two different eggs fuse together into one egg. A pregnant mother can even gain new genomes in her cells from her child, who leaves fetal cells behind after birth. These cells can migrate to the mother's organs, even the brain, and be absorbed. This event is known as mosaicism, and it looks to be far more common than ever imagined. In some cases, mosaicism is believed to contribute to diseases like schizophrenia, but for the most part it is considered benign.

Even among Darwinian strongholds, it's become obvious that evolution is a complex dance between hard and soft inheritance. For example, sexual reproduction in most species is hardwired. A male fruit fly automatically knows that in order to mate, it must find a suitable female, tap her with his forelegs, sing specific songs, vibrate one wing, and lick her genitalia. No one has to teach this to the fruit fly. Every gesture is genetically hardwired, and the program is evolutionarily very old. But at some point long ago, these behaviors were not yet wired in; they had to evolve. Each choreographed component of the mating ritual individually emerged in some ancestral fruit fly male and then began to spread. Eventually the new trait became so successful that mating couldn't take place without it. At that point, we call the ingrained behavior "instinctive," "hardwired," or "genetically determined."

In other words, the behavior occurs with no thought necessary. It arises in response to a specific stimulus. A cockroach will automatically scurry away and hide when a light is switched on. A lizard will scamper off when a person's shadow approaches. A squirrel will enlarge its tail to appear bigger when facing an attacker. These innate behaviors have become automatic to ensure survival. But it goes too far to claim, as evolutionary psychologists do, that human behavior is primarily a matter of survival.

This claim is an attempt to make us seem hardwired like fruit flies, cockroaches, and squirrels. Certainly we've inherited mechanisms from our mammalian ancestors that are innate—the fight-or-flight response is the most obvious example. But we can override our ancestral inheritance at will, which is why, for example, firemen don't run away from a blazing inferno but toward it, or why soldiers on the battlefield will rush in under heavy fire to save a fallen comrade. Mind trumps instinct through choice and free will. In the same way—and this is the idea that outrages mainstream geneticists—mind trumps genes as well.

Is there a survival benefit to art, music, love, truth, philosophy, mathematics, compassion, charity, and almost every other trait that makes us fully human? Are these traits acquired genetically? Elaborate scenarios are devised every day by evolutionary psychologists who insist that they can show why love, for example, is just a survival skill or a tactic that evolved to make mating more possible. Every other trait is "explained" in similar fashion for solely one purpose—to preserve at all cost Darwin's original scheme.

What's anathema is any admission that *Homo sapiens* evolved using the mind, sidestepping genes altogether. Yet at a certain point it's obvious that we pursue music because it's beautiful, practice compassion because our hearts are touched, and so on. In some way these behaviors are inherited, but no one knows how. The existence of mind as a driving force is just as good an explanation as any, and often much better. It's entirely possible that we "download" many of the cherished traits that make us human, not by evolving the tiny gestures that go into a fruit fly's mating ritual, but by taking the whole thing at once.

For example, one hears of a child prodigy who has never had a music lesson and yet instinctively knows how to play an instrument as a toddler. The great Argentinean pianist Martha Argerich relates just such a tale.

> I was at the kindergarten in a competitive program when I was two years and eight months. I was much younger than

> the rest of the children. I had a little friend who was always teasing me; he was five and was always telling me, "You can't do this, you can't do that." And I would always do whatever he said I couldn't.
>
> Once he got the idea of telling me I couldn't play the piano. (Laughter) That's how it started. I still remember it. I immediately got up, went to the piano, and started playing a tune that the teacher was playing all the time. I played the tune by ear and perfectly. The teacher immediately called my mother and they started making a fuss. And it was all because of this boy who said, "You can't play the piano."

It's impossible to know whether Argerich simply inherited either the genes or the epigenetic marks that were responsible for her amazing gift. There are inherited skills. Babies are born with the grasping reflex that allows them to clutch at the breast. They have a sense of balance, and some rudimentary but powerful reflexes for survival. For example, experiments have been done with babies only a few months old in which they are placed on a table while their mothers, standing a few feet away, encourage them to come closer. When the infants approach the edge of the table, they won't go past it; they reflexively know that going over the edge means that they will fall. (There is actually a glass extension to the table, so the experiment is perfectly safe.) Because they want to be with their mothers, the babies start to cry in distress, but no matter how coaxing the mother is, her offspring obeys its innate instinct.

But music is a complex skill involving the higher brain, and unlike a simple reflex, much information must be learned, organized, and stored. How can it be that music prodigies, of which there have been many, somehow inherit a complex mental skill? No one knows, but it argues powerfully for mind being crucial to evolution, since evolution is entirely about inheritance. To deepen our sense of mystery, take the case of Jay Greenberg, a musical prodigy who ranks with the greatest in history, such as Mozart. The first time Jay saw a child-size cello at age two, he took it and started playing. By age ten he entered the Juilliard School with the intention of being a composer, and by his mid-teens Sony had released a CD of his Symphony No. 5, played by the London Symphony, and his String Quintet, played by the Juilliard String Quartet.

As for his working methods, Jay, like many other prodigies, says that he hears the music in his head and writes it down as dictation (Mozart also had this ability, although there is a process of refinement and creativity that goes along with it); perhaps unique to Jay, he can see or hear simultaneous scores in his head at the same time. "My unconscious directs my conscious mind at a mile a minute," he told a *60 Minutes* interviewer.

Prodigies cause amazement, but the whole issue of instinct and genetic memory is an incredibly interesting evolutionary concept. A flatworm can be trained to avoid a light by subjecting it to electric shock whenever it sees it. If the flatworm is then cut in half and the end with the head grows a new tail, or the tail end grows a new head, both halves will continue to avoid the light. How does a newly generated brain retain the same memories as the old one—is memory in this case stored in the DNA of the worm? It's an open question how our own instinctive behaviors became encoded as memories in our DNA. We have yet to discover how long it took for them to be automatically programmed in us.

More interesting, we can ponder which of our behaviors *not* currently programmed or automatic in us right now might become so in the far future. We don't know. But when identical stem cells can become any of two hundred different specialized cells in the body, it is epigenetics and coordinated gene activities that are at play. The highly orchestrated symphonies of gene networks are innate, and they give us the beginning of an answer about how complex skills can be "downloaded" intact. We can't even be sure that *inheritance* is the correct term, given that musical and mathematical prodigies, like genius in general, are just as likely to appear in families with no background in music, math, or high IQs.

YOUR MIND, YOUR EVOLUTION

The purpose of this chapter has been to open new possibilities for you as a person who wants to gain control over your own well-being. We needed to discuss evolution in detail so that you may realize how much control you actually have. Evolving in real time is possible. Let's review why.

Mutations aren't always random but may also be induced by the environment and interactions.

Evolutionary change doesn't need millions of years—it can occur in a single generation (at least in mice and other species).

Genes operate by feedback loops that constantly monitor for new messages, information, and changes in the environment.

The brain constantly interacts with the genome, bringing in the vast potential of the mind to affect every cell in the body.

These four points are the takeaway from this chapter, and they pave the way for the transformation that the super genome facilitates. They also pave the way for transforming our whole notion of how evolution works. You don't need to be concerned with where genetics winds up a generation from now. At the present moment, you have enough knowledge to do something incredibly important—you can cooperate with Nature's infinite creativity.

Evolution, after all, is only a scientific word for the creativity and organizing factors that drive the entire universe, but most especially life on Earth. The super genome records every creative leap that life has taken. Until the appearance of human beings, creatures lacked the self-awareness to examine their evolutionary state. A flatworm that is cut in half and forms a new brain containing its old memories has no idea that this enigmatic event has occurred. But you can use your awareness to direct where your life is going. The super genome will always respond, so even in the absence of rock-solid data, we propose the following possibilities:

> Your intentions have a powerful effect on your genome.
> If you set a goal, your genes will self-organize around your desire and support it.
> Creativity is your natural state—you only need to tap into it.
> You were put here to evolve, and the super genome was put here for the same purpose.

Keeping these conclusions in mind is important, because the environment continues to press new challenges on our genes. Unlike our ancestors, who had to meet pressures from weather and predators, many of these new stresses are unfortunately of our own making: global climate change, increasing pollution, artificially created GMO foods, antibiotic-resistant microbes, increasingly toxic pesticides, and contaminated food and water supplies. We all need to begin arming our genomes to ensure the survival of our species. In other words, we aren't only responsible for our personal health and longevity, which relates to one super genome. The real super genome is planetary, and how you evolve has global implications. We don't pose this as an anxiety-provoking responsibility, but a fascinating challenge. If and when humanity solves these new challenges, it will take a quantum leap in evolution, which is exactly how it has always been and should be.

EPILOGUE

The Real You

If you've ever watched a television show on the Big Bang or a future manned voyage to Mars, you'll recognize a standard moment. Someone stands outside gazing at the night sky and murmurs about what a tiny speck the Earth is in the vastness of creation. We wish that for every moment like this, equal time should be given to William Blake and what he once wrote: "To see a World in a grain of sand, / And a Heaven in a wild flower, / Hold Infinity in the palm of your hand, / And Eternity in an hour. . . ." No one has summarized the story of genetics so succinctly or so beautifully.

A microscopic speck of DNA is the closest we can come to seeing the world compressed into a grain of sand. It defies imagination how Nature devised such a scheme. But it did, and here you are, the expression of that world and the millions of years of evolution that has taken place there. DNA compresses life, time, and space into the same speck. If you reflect on it, this changes everything you know about yourself. At this very moment, you merge with the flow of life as a whole.

The real you isn't bound by limitations, any more than DNA is. How old are you? At the everyday level, you'd count the candles on your last birthday cake. But this excludes the 90 to 100 trillion micro-organisms that are the largest biological part of "you." Single cells can only reproduce by division. One amoeba divides in two, but the two new amoebas aren't its children. They are still itself. In a very real sense, all the amoebas alive today are the first amoeba with select changes in its genome. And, the same goes for all the trillions of micro-organisms that occupy your body and are necessary for it to survive.

Who's the real you? It's the identity you choose to take on. Once you start looking at yourself this way, the individual gradually vanishes. An enlightened Indian sage once told a disciple, "The difference between us can't be seen on the surface. We are two people sitting in a small room waiting for our dinner. But there is still a great difference, because when you look around, you see the walls of this room. When I look around, I see infinity in all directions." If DNA could speak, it would say much the same thing. Time and space are unbounded, and so is the force of evolution that wears human DNA as its crown jewel.

As "you" expand beyond, more and more boundaries can be shed as useless limitations. Since the entire mass of animal and plant life on Earth traces back to single-cell creatures, "you" are one enormous 3.5-billion-year-old being. Separation in space makes each of us think we are individuals. And we are. But the continuum of time at the cellular scale reveals an equal reality: we are united as a single biological being. The human qualities of "you"—awareness, intelligence, creativity, the drive to get more out of life—have a universal source. As we saw, the essentials of human life are present in every cell of the body.

"You" seem to inhabit your body as a life support system of considerable fragility. But even this limit is a matter of what you choose to identify with, the part or the whole. There is no atom in your body that did not derive from something eaten, drunk, or breathed from the substance of the planet. Whether we talk about the "you" that is sitting in a chair reading this sentence or the "you" that is a single enormous 3.5-billion-year-old being, neither lives *on* the planet—they *are* the planet. Your living body is the self-organization of the substance of the Earth itself—minerals, water, and air—into zillions of life-forms. Earth plays Scrabble, forming different words as the genetic letters are recombined. Some words, like *human*, run away to live on their own, forgetting who owns the game.

If "you" are a recreational pastime for the planet, what does it have in mind for its next move? Games involve a lot of repetition, but there has to be novelty as well, with records to break and highest scores to shatter. "You" has its choice of playing fields. At one level, the Mars probe named *Curiosity* can be viewed as a separate human achievement, and a very complex one. It involved skilled, clever engineers and scientists who figured out how to make a robot, propel it to another world, have it land, and then send information back to us. But there's another way of looking at it. Just as reasonably, logically, and scientifically, our living planet is reaching out to touch its neighbor.

The planet has been patient in this endeavor. While "you," sharply focused on the separate self, were busy discovering fire, inventing agriculture, writing sacred texts, making war, having sex, and other survival stratagems, Earth may already have dreamed of tapping Mars on the shoulder. (Rudy is on a task force now aimed at protecting the brains of astronauts from cosmic radiation en route to Mars.) If this image strikes you as fanciful, look at the activity of your brain. You are conscious of having a purpose in mind when you walk, talk, work, and love. But it is undeniable that many brain activities are unconscious, while the activity of the brain as a whole is totally unknown. Whatever makes the Earth a totality makes your brain a totality. Therefore it isn't fanciful to think of the Earth as moving in a coherent, unified direction, just as your brain has from the moment you were born.

Or to put it in a word, if you (as a person) have a purpose, then you (as life on Earth) have a purpose. Perhaps even Earth, as a collection of diverse species, just as we are a collection of microbes and mammalian cells, has a purpose in the solar system, and the solar system in the galaxy, and onward to the universe. Do we, as a species, serve a specific function on Earth, in its capacity as a "being" in the universe? Perhaps we are the immune system of our dear planet. Why? The only natural predator that can turn our planet into a lifeless rock is a giant comet or asteroid. We are the only species on Earth that can predict such an event and have a chance to prevent it. And, like our own immune system, we need it but can also be harmed by it when it goes out of whack—for example, in inflammation and autoimmune disease. These relationships from cells to human, to Earth and beyond, are seamless, even if it suits our pride to stand above and perceive ourselves as entirely separate from our surroundings.

The super genome isn't the end of the story. It's a work in progress. But at the very least it has stitched "you" and all of us into the tapestry of all life and the universe. In an ideal world, this would be enough to save the planet. In healing the environment, "you" would be saving it from destruction. The signs aren't very promising so far. We hope, by offering this book, that the super genome will point more people in the right direction—taking responsibility for our genome and the planet. One thing is certain. Human evolution is mindful, and all that remains is to decide which way its mind will turn—hopefully, it will be toward the light.

APPENDIXES

We've been describing some exciting science in general terms suitable for lay readers. But some readers will have a deeper interest in the underlying genetics. For them, here's some in-depth information about mutations and epigenetic alterations, because the latter are so crucial in pointing the way toward future breakthroughs. In particular, we want to address the common concern over whether "bad genes" destine a person to acquire specific diseases. The answer isn't nearly so simple. But the best clues connecting complex diseases to your genes are based on the science we've been covering. The thread connecting epigenetics and inflammation appears to lead in many directions. It could be the most exciting medical development in decades. Like your genes, inflammation is double edged. Medical science is now unraveling how mechanisms that benefit the body in so many crucial ways somehow can turn on the body and create enormous problems.

These appendixes are devoted to exploring such mysteries.

GENETIC CLUES FOR COMPLEX DISEASES

One result of the advances in genetic technology brought on by the Human Genome Project was next-generation sequencing, which can decipher huge stretches of the genome in short order, so that we can now objectively scan the entire human genome of a patient to find causative mutations underlying their particular disorder. Then it was discovered, as we mentioned before, that for most common diseases with a genetic component, only about 5 percent of the gene mutations associated with the disease are sufficient to cause it. These "fully penetrant" mutations, once inherited, guarantee the disease. (They are also called Mendelian gene mutations, after the famed pea-growing monk Gregor Mendel, the father of genetics.)

In fact, the first Alzheimer's disease genes that Rudy and others discovered in the late 1980s and 1990s contained such mutations. However, in 95 percent of inherited diseases, variations in the DNA of numerous genes (variants) conspire with one another to ultimately determine someone's risk for disease, adding in lifestyle habits and experience. These variants in the DNA are classified as genetic risk factors. While some increase risk, others can protect us from disease. In the majority of cases, however, the outcome depends on environmental exposure and lifestyle.

For a specific individual, discovering exactly how much contribution is being made genetically involves a huge amount of detective work, scouring multiple gene variations at once and comparing the results to the patient's family history, life experiences, and environmental exposures. So despite the considerable success among gene hunters like Rudy and his team, for many disorders—for example, schizophrenia, obesity, bipolar disease, and breast cancer—the gene variants associated with the disease have to date accounted for less than 20 percent of the variance underlying risk.

For most complex diseases, it is now realized, there's an interplay of nature and nurture. In this interplay, the influence of epigenetic factors assumes a major role. Epigenetics mechanisms have already been linked to many diseases, including the childhood disorders Rett syndrome, Prader-Willi syndrome, and Angelman syndrome. In some cases, gene activity is turned off directly by methylation of the bases of DNA in the gene itself. In other cases, chemical modifications (methylation and acetylation) are made to the histone proteins that bind the DNA in order to silence the gene.

But the picture has become still more complicated. Now that we can sequence whole genomes, we are finding that each of us carries up to 300 mutations that lead to the loss of function of specific genes as well as up to 100 variants that have been associated with risk for certain diseases. Moreover, some mutations and DNA variants influencing risk weren't present in the genomes of our parents but occurred anew in the sperm or egg. These are called *de novo*, or novel, mutations. Novel mutations can occur in the sperm and the egg that joined to form your embryo. Such mutations occur 1.2 times every 100,000,000 bases in the two sets of 3 billion DNA bases you inherited from your parents.

That means you harbor in your genome roughly 72 de novo mutations that your parents don't carry in their genomes. (The actual rate of de novo mutation is heavily dependent on the father's age when the baby is conceived. Every sixteen years after the age of thirty, the number of mutations in paternal sperm doubles, which has been shown to contribute to the risk for diseases such as autism.)

In addition to single-base variants in your DNA, you carry large duplications, deletions, inversions, and rearrangements of up to millions of bases of DNA—these are known as structural variants (SVs). Like the single-base variants (technically notated as SNV, for single-nucleotide variant), structural DNA disruptions can either be inherited from your parents or occur as de novo mutations. In Alzheimer's disease, a duplication of the APP (amyloid precursor protein) gene, the first Alzheimer's gene to be discovered, inevitably leads to early-onset (under age sixty) dementia.

SVs and SNVs can both be found by next-generation DNA sequencing. But in another type of genetic analysis, gene expression (or gene activity) can be assessed across the entire genome. This is called transcriptome analysis. When a gene makes a protein, it first makes an RNA transcript that will be used to guide the synthesis of the protein. Transcriptome analysis can be used as part of testing for epigenetic regulation of genes, since it provides information about gene activity, not the sequence of the DNA.

The point is that powerful tools are now available to unravel the complexity of most diseases that have a genetic component. One issue is that the way a complex disease progresses is by a series of steps connected to one another. In everyday life, when you catch a cold, you first notice a mild symptom like a scratchy throat, and unless you catch the cold at this very early stage (by taking zinc tablets, for instance), you know from experience that a chain of symptoms will follow. Something similar is involved in genetics. Genetic studies using transcriptome analysis and whole-genome sequencing together carry out "pathway analysis," which looks at many genes involved with a disease at once. With this information, the aim is to understand the pathological mechanisms by which the disease is caused and progresses. Specific biological pathways—for example, inflammation or wound healing—influence the risk for disease.

Pathway analysis also elucidates other new genes of interest that might be involved in the disease, based on the biological pathways implicated. For example, in Rudy's studies of Alzheimer's disease, pathway analyses of the risk genes that he and others discovered have implicated a major role for the immune system and inflammation. When it comes to human disease, whether it's cancer, diabetes, heart disease, or Alzheimer's, to name a few, inflammation is almost always the killer that takes the patient out. If you wanted to name the epigenetic change that plays the biggest role in modulating a biological process, it would probably be inflammation.

TYPE 2 DIABETES

Close to 400 million people worldwide suffer from type 2 diabetes (T2D), a number that's expected to grow to well over 500 million in the next twenty years. In T2D patients, plasma glucose (or blood sugar) levels are elevated, often later in life as a consequence of both genetics and lifestyle choices, particularly diet. A major risk factor is obesity. One often sees clustering of diabetes in families, and while this would normally implicate gene mutations that run in the family, the family members also tend to eat together, sharing the same diet and probably similar eating habits.

Risk has become more precise but not necessarily simpler. In T2D, dozens of genes are already known to be associated with risk for adult onset. (Not surprisingly, many of these genes have also been associated with obesity and altered glucose levels.) However, most of the DNA variants in the implicated genes exert only small effects on lifelong risk for the disease. Lifestyle is probably most of the story, which you now know means that epigenetics is at work. Some of the strongest evidence for this comes from findings that a person's early diet and nutrition in childhood determine later life risk for diabetes and heart disease. The Pima Indian population in Arizona is heavily affected by T2D and obesity. If a Pima mother was suffering from T2D while pregnant, the children turn out to be highly prone to both T2D and obesity.

The science tying epigenetics to complex disease is emerging at a frenetic pace. We now have gene chip technologies that can search through half a million sites in the genome to find where methylation may be turning off the activity of any of our 23,000 genes. These sites can be scanned for specific diseases like diabetes to ask exactly which genes are being switched. These epigenome-wide association studies, as they are called, are now being carried out around the world for all the most common disorders. In the case of T2D, some of the greatest epigenetic modifications were found around a gene called FTO, which has been linked to obesity and body mass index, which measures the ratio of fat in overall weight.

Another factor contributing to risk for diabetes is birth weight. It turns out that future risk for diabetes is highest in babies born with either low or high birth weights. Epigenetic effects on the genome of low-birth-weight babies can begin in the uterus. For high-birth-weight babies, the issue seems to be exposure to diabetes in the mother during pregnancy. All in all, the risk for T2D almost certainly involves a combination of genes, lifestyle, and epigenetics in which all these factors interplay. The same model is likely to apply for most complex diseases, from metabolic disorders to addictions and psychoses.

ALZHEIMER'S DISEASE

A field of study that has long been close to Rudy's heart is Alzheimer's disease. In 2015 a comprehensive analysis of the role of epigenetics in Alzheimer's was reported in the journal *Nature*, and the results were striking. Researchers at the Massachusetts Institute of Technology (MIT) used mice altered with a human gene that caused them to undergo nerve cell loss, or *neurodegeneration*. This kind of nerve cell death is similar to what happens in the brain of a patient in the final stages of Alzheimer's, which basically robs one of oneself.

As nerve cells started to die in the brains of the mice, the investigators looked for accompanying changes in the epigenome. As rampant neurodegeneration took over the brain, genes in two major categories were found to carry epigenetic marks. These included genes involved in neuroplasticity and the rewiring of neural networks—crucial to the brain's ability to renew itself—along with other genes involved with the brain's immune system. The brain's immune system uses inflammation to protect the brain, often at the expense of nerve cells, which die in the wake of unbridled inflammation.

In the latter case, cells known as microglia, which normally support and clean up after nerve cells, sense the surrounding massacre and assume, mistakenly, that the brain is under attack by bacteria or viruses. Consequently, the hyped-up microglial cells start shooting free radicals (oxygen-based bullets) to kill the foreign invaders. In the process, they kill many more nerve cells as a sort of collateral damage in battle.

The MIT team then compared the epigenomic signature of the brains of the altered mice to the autopsied brains of Alzheimer's patients who had succumbed to the disease. Uncanny matches were observed. (These findings were later extended to epigenetic marks in patients currently suffering from the disease.) Starting in 2008, Rudy's group and others were increasingly finding new Alzheimer's-associated genes functioning as part of the brain's immune system, carrying mutations that predispose to inflammation. When the results of Rudy's Alzheimer's Genome Project were combined with the MIT group's data, the message was loud and clear: Alzheimer's is essentially an immune disease driven by the interplay of immune gene mutations and lifestyle, ultimately culminating in epigenetic alterations of those same immune genes.

An entirely new paradigm for the cause and progression of Alzheimer's disease was being born. Rudy's team and others are still trying to figure out how to "chill out" the brain's immune system as a way to prevent and treat the disease. The answers will undoubtedly lie in the way immune genes are orchestrated to deal with the onslaught of neurodegeneration in the brain.

SLEEP AND ALZHEIMER'S

We'd like to address the intriguing trail of clues that solved one of the chief mysteries behind Alzheimer's disease. As it turns out, sleep was one of the main clues. Disturbances in the sleep/wake cycle have been associated with numerous neurological and psychiatric diseases, including Alzheimer's disease. Science is arriving at a pretty good idea of how sleep is linked to Alzheimer's. We now know that the disorder is initiated by the excessive accumulation in the brain of a small protein called *beta-amyloid*, written variously as β-amyloid and amyloid-β (Aβ), which was not always obvious. When Rudy was a student in the mid-eighties, he and others in the field had maintained that Alzheimer's is initiated by brain amyloid deposits. In 1986, Rudy and others discovered the gene (APP) that makes Aβ (this also turned out to be the first Alzheimer's gene), and twenty-eight years later he and his colleagues developed the first model of Alzheimer's pathology in a laboratory petri dish by growing brain nerve cells in an artificial brain-like environment. In that study, Rudy and colleagues Doo Yeon Kim, Se Hoon Choi, and Dora Kovacs were able for the first time to fully recapitulate the senile (amyloid) plaques and tangles inside the nerve cells that litter the brains of Alzheimer's patients. The study earned the team a highly prestigious Smithsonian American Ingenuity Award in 2015.

The creation of "Alzheimer's-in-a-dish," as the *New York Times* dubbed it when they reported on the scientific paper in *Nature* announcing the achievement, settled a thirty-year debate.* That debate, in fact, had been the biggest in the Alzheimer's field. The

---

*The Alzheimer's-in-a-dish study was made possible by a very forward-looking foundation, the Cure Alzheimer's Fund.

debate was over whether excessive amounts of amyloid surrounding the outside of affected brain cells was the actual cause of forming tangles inside the cells, leading to their death. (Tangles are an abnormal aggregate of proteins inside a brain cell that serve as a critical marker for Alzheimer's.) The new study provided the first convincing evidence that β-amyloid can trigger all the subsequent pathology leading to nerve cell death and Alzheimer's dementia.

Alzheimer's is the most common cause of dementia in elderly people, and sufferers frequently experience major sleep problems. While these sleep disturbances were once dismissed as a simple consequence of the disease, we know they occur early on and may actually help cause Alzheimer's. Considerable evidence indicates that the sleep/wake cycle is tightly linked to the production of β-amyloid in the brains of humans and mouse models of Alzheimer's disease. As shown by Rudy's colleague David Holtzman at Washington University in St. Louis, more amyloid is produced at higher levels in the brain when we are awake and nerve cells are more active. At night, particularly during deep sleep (slow-wave sleep), amyloid production is turned way down. Some other useful things happen in the brain during deep sleep. First, it is believed by some scientists that during deep sleep, short-term memories are consolidated into long-term memories, rather like downloading data from your thumb drive to your hard drive. Second, with regard to Alzheimer's, not only is β-amyloid production turned down during deep sleep, but this is also the time when the brain literally cleans itself out. It produces more fluid around brain cells, which serves to flush out the bulk of metabolites and protein debris like β-amyloid. This waste-clearance pathway is referred to as the brain's *glymphatic system*, resembling what the body's lymphatic system does but employing the brain's glial cells rather than lymph cells. So not only do you get a break from β-amyloid formation as nerve cell activity slows down during deep sleep, but you also clear it out of the brain. Meanwhile, humans or mice that are sleep deprived—a major stressor—make much more β-amyloid and show evidence of elevated nerve cell injury and even tangle pathology. Given that β-amyloid and tangles drive nerve cell death in Alzheimer's disease, there is now an added reason to get eight hours of sleep every night and avoid the stress placed on your system by sleep deprivation. Good sleep is promising as one of the best ways to potentially lower your risk for Alzheimer's. It's also possible that improving the quality and duration of sleep in Alzheimer's patients could help them. While we do not yet understand exactly how sleep cleans out the brain at the level of our genes, attending to your own sleep helps reduce the anxiety provoked by this terrible disease.

BREAST CANCER

Another disease with complex patterns for risk is breast cancer. Researchers at University College London have revealed much of the epigenetic signature for breast cancer by studying healthy women who later went on to get breast cancer, with or without the presence of a mutation in the BRCA1 (pronounced "bra-ca 1") gene. BRCA1 mutations are responsible for about 10 percent of breast cancers, leaving the other 90 percent largely a mystery. The question is, how much "missing heritability" is epigenetic? It turned out that the epigenetic alterations involved were pretty similar in both groups of women; in other words, the alterations were independent of inheriting the BRCA1 gene mutation. If the disease's epigenetic signature is known, it can eventually be used to predict who is on the way to getting breast cancer before it strikes, a major advance given that every year 250,000 women get the disease, and 40,000 die from it.

The fact that epigenetics has such an apparently strong effect on risk means we must deeply consider lifestyle changes, beginning with diet. Among nutrients and supplements that have been validated to help reduce the risk for breast cancer are aspirin, coffee, green tea, and vitamin D.

In the case of aspirin, the best data come from a 30-year study that followed 130,000 people. Those who regularly took aspirin (at least two 325-milligram aspirin tablets per week) had a decrease in gastrointestinal cancer of 20 percent and a decrease in colorectal cancer of 25 percent. The results for these specific cancers don't apply across the board to cancer in general, and it took 16 years of taking aspirin for the benefit to appear. If people stopped taking aspirin for 3 or 4 years, their advantage disappeared. The reason aspirin works against cancer, so far as is known, is connected to its anti-inflammatory effect (no surprise) and its apparent ability to decrease the formation of new cancer cells.

HEART DISEASE

In heart disease, we also know that gene mutations and lifestyle work together to determine risk, but as in diabetes and breast cancer, so do epigenetic modifications (methylation) that silence certain genes. In one study it was found that levels of two blood fats (triglycerides and very-low-density lipoprotein [VLDL] cholesterol) were tied to methylation of a gene called carnitine palmitoyltransferase 1A (CPT1A). This gene makes an enzyme needed to break down fats. When it is turned off by epigenetic mechanisms, instead of fatty acids in the body being converted into energy, they stay around in the bloodstream, increasing the risk for heart disease. Methylation of the CPT1A gene is affected by diet, alcohol, and smoking.

ALCOHOL AND GENES

Even alcohol dependence is affected by epigenetic events. Alcoholism takes a devastating toll on the victims as well as their families, contributing to 1 in 30 deaths worldwide. The best-known genes associated with alcohol dependence are alcohol dehydrogenase (ADH) and aldehyde dehydrogenase (ALDH). Both make enzymes that help break down alcohol in the body. But variations in these genes explain only a minor degree of the inheritability of alcoholism. The "missing heritability" likely lies in epigenetic changes that are tied to the reward centers of the brain, the source of feeling good when you take a drink.

Now we know that these reward centers actually undergo changes in gene activity following the intake of alcohol. This means that different people will respond to alcohol consumption in different ways, depending on their gene activities. In heavy drinkers, an amino acid called homocysteine may go up, ultimately leading to methylation changes that silence specific genes. Such gene activities can trigger a vicious circle in which the response to pleasure and pain is altered, leading to an increased craving for alcohol to deliver less and less pleasure.

MENTAL ILLNESS

Epigenetic modifications can also be tied to psychiatric disorders like schizophrenia and bipolar disease. Finding the inherited gene mutations that lead to these illnesses has so far met with only limited success. This impasse once again leaves a potentially significant role for epigenetics in helping to fill in the missing heritability and the role of lifestyle. Increasing evidence shows that schizophrenia and bipolar disorder may not be guaranteed by, or solely dependent on, gene mutations that are passed on from parent to child.

Suspected culprits in someone's lifestyle include diet, chemical toxins, and child rearing that affects epigenetic modifications. A patient's lifestyle can determine epigenetic marks acquired since birth, but mouse studies would suggest that other epigenetic marks may be inherited. These marks presumably would arise as a result of the lifestyles of the parents or even grandparents. (Please note that we aren't suggesting blame. The epigenetics of mental illness are quite tentative and incomplete. No one has yet connected A to B for any lifestyle choice that may be implicated in mental disorders.)

Epigenome-wide studies of schizophrenia and bipolar disorder have revealed epigenetic marks on some predictable genes, such as those involved with making certain neurochemicals previously associated with psychosis. But others were less predictable. For example, key genes required for immunity have turned up in both schizophrenia and bipolar disorder, suggesting that the immune system may be somehow related to a susceptibility to these disorders. Of course, here and in other epigenetic signatures associated with risk, cause and effect are an issue. How do we know whether the epigenetic marks occurred previous to onset (cause) or as a result of the disease (effect)? For now, it's safe to say that epigenomic tests for specific diseases will become invaluable in every aspect of preventing and treating complex diseases, from prevention to ultimate cure.

In fact, we are tremendously optimistic about where genetics is leading, but we are realists, too. There remains a sharp divide between two domains, the visible and the invisible. All of us live in both domains, a fact that can't be ignored. Peering through a microscope, a cell biologist can witness myriad changes in how a cell is functioning, yet the most crucial component, the experience that guides these changes, cannot be observed. The nonphysical is playing its part during every second of a person's life, and we believe it's the prime reason why genetics must look beyond materialism and random chance.

The data will have to support such a radical change in perspective, but far more important is to formulate the ideas that the data must fit—that's our objective in this book, and we've taken some giant steps in that direction. You now know more about the dynamic nature of your genome than geneticists knew even twenty or thirty years ago. What's most crucial, however, is applying the knowledge to optimize your genetic activity. Before we can do that, another big chunk of genetic information needs to be presented, and it comes from a very surprising source that no one ever anticipated.

THE GREAT PARADOX OF DNA

Epigenetics is a complex subject, and in reading this book you've grasped the main concept: that gene expression is switched on and off and up and down based on the choices you make every day and the resulting experiences that create who you are. This switching, which leads to trillions and trillions of possible combinations, is how everyday experience is transmitted to the cells of your body. But immediately a troubling problem arises. Why are some experiences so damaging to the body? Why isn't DNA designed to preserve life as its only mission?

This is the great paradox of DNA, and it forms the next link in our story. DNA makes life possible, but at the same time it has the potential for ruinous, life-destroying actions. DNA is like a bomb that knows how to defuse itself and also how to set off an explosion. Which one will it choose? Why should the code of life be employed to create death? That's the heart of the paradox. In all of us there are genes for developing cancer (pre-oncogenes) and opposite genes for fighting cancer (tumor-suppressing genes). This seems inexplicable until you realize that DNA reflects every aspect of existence.

Instead of choosing sides, DNA joins all sides, encompassing all possibilities. A virus or bacterium that can make you sick has its own genetic signature, which it does everything to keep intact, and so do the immune cells in your body that war against viruses and bacteria. When new cells are born, they inherit a genetic program for their death. In effect, DNA is staging a drama in which it plays the role of hero and villain, attacker and defender, keeper of life and destroyer of life.

The challenge is to make choices that activate the life-supporting side of DNA. By now you've seen that we've taken big steps in that direction. You've started to view life from the perspective of a cell. A cell senses its environment and makes adaptations that best serve its own survival. But it also does this using the least possible energy to maintain balance and serve its neighboring cells and the whole body. Failure to do so can lead to cancer or other diseases that can potentially kill the host and the cell along with it. So every cell naturally knows exactly what to do in all situations working in perfect harmony with its genes. Our hope is that we can do the same as human beings.

The latest research into a wide range of disorders, including heart disease, autism, schizophrenia, obesity, and Alzheimer's, suggests that there are indicators for each disease that extend back decades in a person's life, even to early infancy. This came as a startling discovery, because it runs contrary to our conventional notion of how we get sick. We tend to believe that getting sick follows the pattern of the common cold. You are sitting on a plane next to someone who is sneezing and coughing. Three days later you catch that person's cold. There is a simple cause-and-effect, along with a definite starting point of infection.

Many acute illnesses do in fact follow this pattern, but it turns out that chronic disease doesn't, and chronic disorders are the major causes of mortality in modern society. How do you organize a prevention program for a disorder decades before symptoms appear? A perplexing example of this dilemma actually showed up in the Korean War, when autopsies were performed on the bodies of young soldiers killed in battle. Males in their early twenties exhibited the fatty plaque in their coronary arteries that are the major cause of heart attacks. How did men so young have this much plaque, often enough to cause worry about an impending heart attack? There was no medical answer, and even today, the genesis of arterial plaque remains to be explained. Just as baffling, why didn't these men suffer heart attacks at a young age, since the onset of premature heart attacks typically begins at forty? Even without satisfactory answers, here was an early clue, going back to the fifties, that chronic disease predates the arrival of symptoms by many years and has no definite beginning except at a microscopic level.

But there's also a very hopeful side to the mystery. These early indicators hold out the best chance for preventing and curing chronic illness, because whenever the body goes out of balance, the earlier it's caught, the easier it is to treat. Millions of people follow this principle when they take zinc tablets at the first sign of a cold or aspirin at the first hint of a headache. The same principle can be pushed back even further, which is why vaccines are effective. They give the body an advance defense against polio, measles, or this year's flu before the disease has had a chance to develop.

In effect, a vaccine is teaching the body's intelligence something new. The body listens (i.e., the genes respond in a new way) and learns from the new experience. "This is what measles looks like. Arm yourself." There's never going to be a universal vaccine for all human ills (even current vaccines have their critics and problems). Instead, we are proposing a new model for self-care; at the heart of this model is a revolutionary way of relating to your genes.

This shift in thinking agrees with every advanced trend in medicine, but the general public hasn't absorbed as yet how radical the change will be. A new era in well-being is at hand, looking to the body's intelligence as our most powerful ally.

To show why this approach is so urgently needed, let's look at a dreaded disease in order to make a much bigger and more optimistic point about well-being. The disease is lung cancer. The war against lung cancer poses a stark confrontation between smoking on one hand and prevention on the other. The battle lines could hardly be clearer. Lung cancer is the leading cancer killer among both men and women, outstripping the next three cancers combined (breast, colon, and pancreatic). It surprises most people to discover that as far back as 1987, lung cancer surpassed breast cancer as the leading cause of cancer deaths among women.

The disorder would be rare if it weren't for tobacco. In 1900, before the general spread of smoking, cases of lung cancer were so uncommon that a doctor in general practice might know of the disease only from textbooks. With the dramatic rise of smoking in modern times, tobacco-related lung cancer accounts for 90 percent of cases, and when someone stops smoking, the risks decrease year by year, although they never reach zero.

Those are the statistics (as provided by the American Lung Association), and ever since the Surgeon General forced tobacco companies to print a warning on every pack of cigarettes in 1964, sensible prevention has been clear and undeniable. (The sad fact that more women today choose to take up smoking is why lung cancer has increased among women.)

But here is where the dividing line between well-being and radical well-being shows up. The fact is, not all smokers contract lung cancer. Why not? The pathogens in tobacco smoke are almost guaranteed to damage lung tissue. A host of respiratory problems, including emphysema and asthma, loom for active smokers. Yet consider the statistics cited at http://lungcancer.about.com.

In a 2006 European study, the risk of developing lung cancer was:

0.2 percent for men who never smoked (0.4 percent for women)
5.5 percent for male former smokers (2.6 percent for women)

15.9 percent for current male smokers (9.5 percent for women)
24.4 percent for male "heavy smokers" defined as smoking more than 5 cigarettes per day (18.5 percent for women)

An earlier Canadian study quoted the lifetime risk for male smokers at 17.2 percent (11.6 percent for women) versus only 1.3 percent in male nonsmokers (1.4 percent in female nonsmokers).

These percentages translate into a story line. If you don't smoke, lung cancer is very unlikely to strike you. If you take up smoking, the odds against you increase in a straight line. However, even if you fall into the highest risk category of "heavy smokers," 75 percent of the time you won't contract lung cancer.

We aren't remotely suggesting that you take your chances and start smoking. The story line actually leads in a very different, and unexpected, direction. Why do some smokers dodge the bullet? This is the million-dollar question that statistics do not readily address. What you and I and every other individual want to know is how our situation will turn out. Lung cancer is only one horrible example. The statistics around every disease point to some people who manage to escape getting ill. "How do I become one of those people?" is the question that naturally arises.

The answer is genetic, but it goes far beyond the cliché that some people have good genes and some people have bad genes. Imagine tobacco smoke entering the lungs of two people. The toxic chemicals in the smoke are the same for both; the known carcinogens are the same. When the smoke hits the outer lining of lung tissue, damage is bound to occur—but not necessarily in the same way or to the same degree.

Cells are very resilient, and they make choices all the time. Over millions of years of evolution, one choice stands out. Cells choose to fight back against any threat to their survival. A major threat, and the one that applies to tobacco smoke, is deleterious variants that arise in genes called pathogenic mutations. The toxins in tobacco smoke can cause a sudden mutation that leads to a distortion in how the cell operates. But DNA knows how to regulate and repair itself, and the norm is for damaging mutations to be destroyed. *There's a limit to a cell's healing abilities, but the cell isn't simply poisoned to death.* With enough exposure to the toxins in tobacco, some distortions will inevitably get by the cell's defenses, and if enough damage occurs, and if the damage is of a precise kind, disaster follows. The cell forgets how to divide normally. A cell that goes on the path of rampant division, overwhelming adjacent cells in its unregulated growth, has become cancerous.

You can see where the story line has now taken us. Behind the statistics for the whole population, the beginning of a malignancy is about single cells deciding what to do, guided by their DNA. Let's press the investigation further. When three out of four heavy smokers escape lung cancer (by no means are they guaranteed to escape other serious illness), what choices did their cells make? For it's those choices that actually rescued them.

The best medical knowledge has this to say: Some people are better at fending off toxins than others. Some DNA is better at repairing itself and destroying harmful mutations. Many factors are at work in how a cell heals, and its escape from danger is blurred into everything else that's happening to it. When it comes to a cell and how it escapes disease, there's a lot of room for uncertainty. Knowing how a typical cell makes decisions doesn't tell us how *your* cells make decisions. Everyone's cells are different, based on their specific component of genes and the gene activities you impart to them with your lifestyle. There's also the whole issue of the paths your cells will make a day, a month, or ten years from now, because like people, cells can be fickle and changeable, depending in part on the choices you make.

We've been dwelling on a grim subject in order to shed light on something positive, the enormous intelligence and resilience of the cell, meaning *your* cells. Research has shown that thousands of potentially damaging abnormalities are detected and destroyed in our bodies every day. What makes the difference between well-being and radical well-being is *learning to guide and influence your genes in a positive manner.*

We said that you are more than your genes, just as you are more than your brain. You are the user of your genes and your brain. The key is learning how to use them so that they afford you optimal health and happiness. Everything you want to be, every achievement you want to reach, every value you want to uphold must pass through your brain and your genes in order to become real. So learning to communicate with your genes isn't just a nice add-on. It's essential. You are already communicating with your genes, but most of the messages you're sending to them are unconscious. Repetition plays a large part. Reactions become automatic and ingrained. This is a terrible waste of your potential to make free choices.

IS DEPRESSION GENETIC?

Genetics would be much simpler if it traveled down a one-way street where gene A could always be connected to disorder B. *Linear cause-and-effect is* simple and satisfying. But genes operate on a two-way street, with messages constantly traveling back and forth—or, to be more accurate, the road is a six-lane superhighway, loaded with messages coming from all directions.

This realization is having a huge ripple effect throughout medicine and biology, overturning what we thought we knew about the brain, the life of a cell, and almost every form of disease. To give a prime example, we'll look at the present situation in depression, which directly or indirectly has touched almost everyone's life, either through their own suffering or that of a family member or friend.

About 20 percent of people will experience a severe depression sometime in their lives. At the moment, there is a rash of depression among combat soldiers who served in Afghanistan (directly related to a sharp increase in suicides among Afghanistan war veterans, suicide being generally linked to depression) and among laid-off workers who are enduring long-term unemployment. In both cases, an outside event led to the depression, but we do not know why, in the sense that only a certain percentage of people become depressed under the same stimulus (war and losing a job).

The link between depression and genes has proved elusive. Nothing as simple as a "depression gene" exists. Early in 2013, the magazine *Science News* began an article on depression with a blanket judgment: "A massive effort to uncover the genes involved in depression has largely failed." This news sent shock waves through the medical community, but its impact hasn't really hit the public, which keeps funding the multibillion-dollar drug industry and its constant production of new—and supposedly better—antidepressants. Twenty-seven years after Prozac came on the market in 1988, around one in five Americans takes a psychotropic (mind-altering) drug, despite the proven risk of side effects. Prozac, for example, has three common side effects (hives or skin rash, restlessness, and the inability to sit still); two less common ones (chills or fever and joint or muscle pain); and twenty-five rare ones (including anxiety, fatigue, and increased thirst), according to the website www.drugs.com.

The connection to genes isn't brought up when the physician is prescribing a drug to relieve a patient's suffering. However, genes are the pivot between a drug that works and one that doesn't. The model for depression that has been accepted for decades labels depression as a brain disorder. Yet brain disorders are rooted in genetics. The logic is deceptively simple. If you feel depressed, there is an imbalance in the brain chemicals responsible for moods (chiefly the neurotransmitters serotonin and dopamine). Thus in depression, the cellular mechanism that produces these chemicals must be impaired, which comes down to impaired genes, since genes are the starting point for every process taking place inside a cell.

Why didn't this simple logic turn out to be true? As prominent researchers now concede, the genes of depressed people are not damaged or distorted compared with the genes of people who aren't depressed. What follows from this finding is that other basic assumptions are wrong. The most popular antidepressants supposedly worked by repairing chemical imbalances in the synapses—the gaps between two nerve endings—where the culprit was an imbalance of serotonin. But serotonin is directly regulated by genes, and some key research indicates that either drugs aimed at fixing the serotonin problem don't work that way or there wasn't a serotonin problem in the first place. The *Science News* report didn't leave much wiggle room on this point: "By combing through the DNA of 34,549 volunteers, an international team of 86 scientists hoped to uncover genetic influences that affect a person's vulnerability to depression. But the analysis turned up nothing." (The study being referenced was published in the January 3, 2013, issue of *Biological Psychiatry*.)

Nothing doesn't mean something. If the chain of explanation running from genes to the synapses and finally to the pharmaceutical lab is broken, a host of doubts arises. Is depression a brain disease in the first place, or is it, as psychiatry assumed before the arrival of modern drug treatment, a disorder of the mind? The latest theories haven't gone back to square one. What we know isn't black and white. There are multiple variables in depression, which leads to some fairly good conclusions:

There are many kinds of depression. It isn't a single disorder.

Each depressed person displays their own mixture of possible causes for their symptoms.

The mental component in depression includes upbringing, learned behavior, core beliefs, and judgment about the self.

The brain component includes wired-in neural pathways,

with suggested weaknesses in certain areas of the brain whose cause isn't understood.

Depression can't be isolated to one region of the brain. The interaction of multiple regions is involved.

As you can see, these conclusions scuttle a simple cause-and-effect model. "If you have a headache, take an aspirin" doesn't translate into "If you feel depressed, take an antidepressant." The susceptibility to depression is as complex as gene expression itself. Why does depression run in families, as it's known to do? Again, there's no simple answer. No gene or group of genes that you inherited seems to guarantee that you will become depressed. We are talking instead about genes that make you susceptible to the disorder. What triggers these (unknown) genes remains a mystery. The same genetic predisposition could be hidden in one child who never becomes depressed when he grows up and in another child who somehow gets triggered into depression. Do social interactions, for example, make someone feel helpless and hopeless? That's how depression feels, so perhaps (in the epigenome) enough bad memories of feeling left out or ostracized from others lead to a tipping point and depression emerges.

In our opinion, depression isn't a brain disorder looking for a magic bullet to solve it, and the whole disease model must be drastically changed. Even as a medical diagnosis, it's suspect. The big study about the failure to find the genes responsible for depression ignored diagnoses of depression and went with symptoms instead. Asking people about their symptoms resulted in a lower number of those who would be considered depressed. Perhaps some people are in denial or don't know the difference between depression and ordinary sadness. But more important, symptoms change over a lifetime, and there is a sliding scale for each sufferer. Like emotion in general, depression comes and goes. It feels different one day than it does another.

So will depression ever be curable? The situation is too cloudy for anyone to offer either a pessimistic or optimistic prediction. Drug treatment remains hugely popular, no matter what the basic science says. In cases of mild to moderate depression—the most common type—antidepressants sometimes don't work better than 30 percent of the time, around the same as the placebo effect. Some symptoms of severe depression remain intractable, and yet in other cases, the chronically depressed perform the best with drug treatment. Hope is always better than giving up.

Now that you understand the situation, with all its uncertainties, you are ahead of the curve, because the vast majority of doctors turn their back on the research and keep prescribing the same antidepressants. Millions of patients continue to take them, feeling that there is no other way. But there is. Depression doesn't fit the old disease model, but it does fit the new model we've been describing. Depression involves lifestyle and environment. Genes play a part but so do behavior, beliefs, and how a person reacts to everyday experiences. The epigenome is storing genetic reactions of personal experiences and memories, leading to the constantly shifting activities of your genes.

EPIGENETICS AND CANCER

Let's expand on what is known about genes and cancer. Perhaps no disease relies more on genome-related risks than cancer. To explain why, we need to backtrack a moment. As mentioned earlier, while still a student at Harvard Medical School, Rudy was thrilled to participate in the first study to ever find the gene for a disorder of unknown cause (Huntington's disease). Since those pioneering studies using genetic analyses in the early 1980s, the hope has been that all of the mysteries of inherited disease could be solved by comparing the genome of patients versus that of their healthy counterparts. In that total of 6 billion letters, combining A, G, C, and T, inherited from our parents, only about 200 million are used to make up the genes. The sparsely distributed genes are like words in the story of life told by the genome. The remaining 5.8 billion letters serve to arrange and punctuate those words, potentially creating many variations of the same story. For the most part, after the Huntington's disease gene discovery, from 1990 to 2010, geneticists spent most of their time looking for disease mutations only in the DNA sequence of the genes, like typos in the words of the genome story. But epigenetics now tells us that much of the story is in that intergenic DNA, the regions of the genome that we used to call "junk DNA" lying in between the genes. These regions determine how the story is read and which chapters matter most.

In an editorial in *Nature* accompanying the first data to emerge from the comprehensive catalog known as the Roadmap Epigenome Project, it was stated: "In human diseases, the genome and epigenome operate together. Tackling disease using information on the genome alone has been like trying to work with one hand tied behind [one's] back. The new trove of epigenomic data frees the other hand. It will not provide all the answers. But it could help researchers decide which questions to ask." It turns out that most common diseases with a genetic basis are highly complex, and a large number of factors, ranging from genome mutations inherited from our parents to epigenetic modifications brought on by life experiences, conspire together to determine one's risk for specific diseases.

In the decades-long war on cancer, definite progress has certainly been made. But according to the American Cancer Society, as of 2015, over 1.6 million Americans are still diagnosed with cancer each year and nearly 700,000 succumb to cancers of all types. More than any other disease, cancer has led to incredible progress in understanding the genetic mutations responsible for the disorder. And, the current belief is that the development of cancer is due to the accumulation of gene mutations causing the cells to become cancerous and form tumors of various types. However, we now know that the risk for cancer is also dependent on the way in which epigenetic modifications to the genome render certain regions more prone to newly occurring mutations. (To date, the greatest evidence for the role of epigenetics in disease comes from cancer studies, in fact.) These mutations can be triggered by exposure to certain environmental toxins—for example, dioxin, a lethal family of chemicals found in pesticide manufacture and waste incineration, for which there is no safe dosage. The Environmental Protection Agency estimates that the damage being caused by dioxins outstrips that caused by DDT in the sixties. An environmental toxin can have the ability to cause new epigenetic alterations. These can modify how the genomic DNA in that region is folded, which in turn can potentially affect where new mutations are allowed to form.

Thus tumor formation involves multiple steps including both genetic and epigenetic alterations in the genome. Unlike gene mutations, the epigenetic modifications can be considered impermanent and even reversible. Some forms of cancer are brought on by genes that are activated via a process called hypomethylation (*hypo* is a Greek prefix meaning "under"). In this case the methyl marks on genes that silence their activity have somehow become removed. Without a suppressor to hold them back, the harmful genes are activated. In other cases, the reverse happens. Turning *off* certain genes via methylation can lead to tumor formation or can involve the addition of acetyl chemical groups to the histone proteins that wrap around the DNA.

New drugs are now being developed that would offset these tumor-causing epigenetic alterations. For example, drugs known as DNA methyltransferase inhibitors (DNMTIs) act as demethylating agents that can remove methyl marks from genes. Such drugs are already used successfully to treat forms of leukemia. Other drugs, called histone acetylase (HDAC) inhibitors, are also being used for treatment of leukemia and lymphoma. Of course, these so-called epidrugs are not without problems, since they are terribly specific in their actions on the genome. And while they are being used with some success in treating blood cancers, they have not yet been very effective against solid tumors. While we hope for the best with this new class of epidrugs, we must also consider the need for studies of lifestyle changes—for example, healthy diet, stress management, exercise, weight control, and the like—that would achieve the same outcomes.

IS CANCER RANDOM?

Randomness is more than a theoretical issue—in our own lives cancer causes a major portion of human suffering. Twenty years ago, in the 1990s, it was thought that cancer was essentially random, putting almost everyone at equal risk. Genetics reinforced the public image of cancer as ruthlessly impersonal, striking any victim it chose. There were countering arguments. Those who thought that cancer was caused by toxins pointed to tobacco and asbestos as prime examples. Others who argued for viruses pointed to cervical cancer, which is caused by the human papillomavirus (HPV). It turned out that everyone had a piece of the puzzle, or as one leading cancer expert called it, each camp was like one of the blind men holding on to a different part of the answer.

The current view brings us back to our familiar image, the cloud of causes. Environmental toxins, viruses, and random mutations all play a role, and as with the puzzle of why Dutch men suddenly became the tallest in the world, the cloud isn't very satisfactory when trying to link cause and effect. The only real certainty is that all roads lead eventually to the genome. Cancer of any kind is now known to need a trigger inside the cell, in the form of a cancer gene (oncogene). There are many such genes, and in recent years they've been cataloged by a worldwide effort to formulate the Cancer Atlas, a complete genetic road map to the disease. Besides turning on an oncogene, cancer can begin by turning off its opposite, the tumor suppression gene.

Once one talks about switches being turned on and off, epigenetics enters the equation, and so do questions concerning randomness, because the event that triggers the switch may not be random at all. Smoking cigarettes isn't a random event. If you smoke, your risk of contracting lung cancer enters the realm of high probability. But the epigenetic explanation for cancer offers as many problems as solutions. For one thing, the futile hope that cancer might involve a single gene, which perished three decades ago in the 1980s, has been repeated in epigenetics—it turns out that while one gene mutation may lead to a certain type of cancer, the disease seems to involve up to *fifty or one hundred genes.* Cancer genes can continue to mutate as the cancer spreads, making the malignancy a fast-moving, extremely elusive target. Gene-targeted drugs have garnered headlines by curing specific cancers like one form of childhood leukemia that involves only a single gene.

After two decades of searching for similar drugs to wipe out a variety of cancers, however, success has been very limited. To make matters worse, drugs that work brilliantly at wiping out all traces of malignancy often have a tragically temporary effect. The patient returns after a few months with his cancer returned. On the surface, it would seem that cancer's secret weapon is how quickly and randomly it can mutate, upholding the evolutionary dogma that randomness rules.

But there are signs pointing in a new direction. Of all diseases, none more than cancer has been clearly linked to epigenetic aberrations.

The epigenomes of specific types of cancer cells carry the same epigenetic fingerprint that matches the cell that started the cancer. This serves to reveal the tissue in which the cancer originated, no matter where in the body it is found. Such information could be of immense use in the future for diagnosing and treating different forms of cancer, because once it has spread, a tumor has often been extremely difficult to trace back to where it started. Further complicating the problem is the cancer cell's habit of continual mutation. Hopefully, by comparing the epigenomes of healthy and malignant cells, we can better understand how the risk for disease can be influenced by much more than the genomes provided to us by our parents.

It turns out that carefully examining the epigenetic marks (methylation and acetylation) can actually be predictive of which kind of cancer will develop. *This* revelation turns out to be the opening wedge against random mutations. As you live your life, and your environment and experiences chemically govern your gene activities—we've discussed this extensively already—specific new mutations can arise that are the same for every cell in a particular type of tumor. So epigenetic modifications lead to *predictable* new mutations. Something that's predictable steps away from being purely random.

This level of predictability doesn't solve the entire mystery, however. By analogy, think of the weather. On a summer day in August, thunderstorms are very likely to arise, and their timing can be predicted with a fair degree of accuracy—as the heat of the day builds up, a storm is more likely in the afternoon or evening than in the cool of the morning. But the exact movement of air currents, moisture, and clouds is much less predictable, and if you want to know the cause of a specific thunderstorm down to the last molecule of air, it's impossible. In cancer, many mutations often occur simultaneously, and not all lead to bad results. Thousands of possibilities arise, with great unpredictability. (Because something is unpredictable doesn't make it random. The next thought you are going to have isn't random, but it is unpredictable. Cancer research has yet to figure out if cancer is like that or not.)

This realization created immense discouragement following hard on the triumphant findings about the genetic causes of cancer. Oncologists began to mutter about cancer as a devious enemy whose arsenal of defenses kept increasing every time a solution seemed to be at hand (a good example of our point in the previous chapter that cancer unfortunately can draw upon the cell's complete intelligence). Now hope is rising again, because the Cancer Atlas has been sorting out which mutations are the dangerous ones, but just as important—and perhaps the single best clue to curing the disease—it appears that cancer develops along some set pathways that are fairly small in number, perhaps only a dozen for every kind of malignancy. In other words, there's a pattern that goes even further to undercut the orthodox view about random mutations.

One promising finding is that certain tumors take many years, even decades, to develop after the initial trigger starts a cell on an abnormal course. The thought is that a specific sequence—the genetic pathway that an abnormal cell must follow—involves a series of steps that must unfold in order. Here's an analogy: You've probably seen the little handheld games that involve steel BBs rolling around on a board with holes in it, the object of the game being to tilt the board around until you manage to get all the BBs to fall through a hole. The holes are tiny, so it's not an easy challenge. Now imagine that a cancer mutation is faced with a similar challenge. It must thread its way through a small opening (a specific genetic modification out of myriad possibilities) in order to move on to the next stage. Once that's accomplished, the next small opening presents itself in the form of a new mutation out of myriad choices, and so on.

If a cancer is typically slow growing, as types of colon and prostate cancer are, it may take thirty or forty years for a cancer cell to follow the whole sequence. The hope is that if detection can be made as early as possible—detecting the predictable fingerprint of epigenetic markings—cancer will be conquered long before the first symptoms appear. This glimpse of light at the end of the tunnel follows from the discovery that the exact gene mutations of many types of tumors can now be predicted from the epigenomic signature of the cell type from which that cancer most likely originated.

We must then at least wonder, is it possible that when epigenetic mutations arise in adults as a result of toxins, stress, trauma, diet, and the like, predictable new mutations will arise in certain cells? If the mutation occurs in sperm and egg cells, could they be passed on to the next generation? We don't yet know. But even the possibility would have made Darwin's head spin and is today leading to a major revision of his theory.

If epigenetic alterations do lead to specific mutations beyond those that cause tumors, then one's life experiences and environment could, at least theoretically, lead to expanded predictability. There could be epigenetic signatures of other chronic illnesses that appear long before the first symptoms. It would be even more amazing if prevention extended to unborn generations that have been inheriting these marks in the womb. At the time this book was being written, such possibilities were only a very intriguing set of conjectures. Yet it's fascinating to think about what future studies in this area will reveal.

ENVIRONMENTAL TOXINS AND EPIGENETICS

So far we've been focusing on the genetic contributions to disease risk, but there's an elephant in the room—the impact of environmental toxins on our genes and epigenome. The Centers for Disease Control and Prevention has found 148 different environmental chemicals in the blood and urine of the U.S. population. Increasing evidence gives support to the notion that environmental pollutants likely cause various diseases by inducing epigenetic changes in our genome, thus altering the activities of specific genes. For example, arsenic in contaminated water dramatically affects methylation of the genome, leading to bladder tumors. Exposure to high levels of other heavy metals (nickel, mercury, chromium, lead, and cadmium) in food and water supplies can also cause changes in gene methylation, leading to various types of cancer, including lung and liver cancers. The bottom line is that there are an estimated 13 million deaths or more worldwide due to environmental pollutants, many of which have been linked to epigenetic modifications of the genome.

We are not alarmists, but it's important to follow where the science leads. Perhaps no one has advanced our knowledge of this issue as much as Dr. Michael Skinner, a developmental biologist at Washington State University. In one study Skinner exposed pregnant rats to a chemical known to interfere with embryonic development, a fungicide called vinclozolin used to keep mold off vineyard grapes, along with other blights and rots on fruits and vegetables. Vinclozolin had already been shown to decrease fertility in male mice. The disturbing thing that Skinner found was that the progeny of the chemically treated mice, all the way down to the fourth or fifth generation, were also affected with low sperm counts. This result was successfully replicated fifteen times.

The reason for the disruption of sperm production brought on by vinclozolin wasn't mutations in the DNA, but epigenetic modifications in the exposed adult mice (via methyl marks), which were then passed on to the next generations. (*This* is different from what we normally hear about, when actual mutated genes for disorders get passed on from parents to children, as in sickle cell anemia.) Thus another clue was being added to the existence of "transgenerational genetics."

Moreover, Skinner and his colleagues found that there was a specific pattern to where the methyl marks were attached in the genome after exposing mice to different types of chemical toxins. Each toxin, whether it was insecticide or jet fuel, left its own distinctive pattern. In some cases, the shifts being caused in gene activity could then be inherited and predispose the offspring to specific disorders. For example, the insecticide DDT, which has long been banned in the United States because of its disastrous effects in the food chain of animals and birds, also has a specific epigenetic effect. Exposing mice to DDT has been shown to create a predisposition to obesity in later generations, along with obesity-associated diseases such as diabetes and heart disease.

The range of detrimental epigenetic changes brought on by pesticides is wide. The pesticide methoxychlor, used to protect livestock from fleas, mosquitoes, and other insects, has been shown to cause testicular and ovarian dysfunction in mice. Another pesticide, dieldrin, has dramatic effects on epigenetic modifications (acetylation) to histones leading in mice to nerve cell death associated with Parkinson's disease. Skinner also showed in mouse studies that the common pollutant and carcinogen dioxin, a waste product of many industrial processes, causes epigenetic inheritance of prostate disease, kidney disease, and polycystic ovary disease.

One of the most carefully studied environmental toxins that can cause abnormal epigenetic changes is bisphenol, or BPA. It has been widely used to make the plastics used in food and beverage containers, including baby bottles. BPA is well known to cause epigenetic changes. We'll cite just a sample of relevant studies. Research at Tufts University showed that BPA can change gene activity in mammary glands of rats exposed to the chemical in the womb, rendering them more vulnerable to breast cancer later in life. Previously BPA was demonstrated to leave male rats at higher risk for prostate cancer. In another set of studies, BPA produced epigenetic changes associated with changing the yellow color of a particular breed of mouse as well as increasing the risk for cancer. (Note: One way to avoid BPA exposure among infants is to use glass bottles and containers or look for the label "BPA free.")

Finally, diethylstilbestrol (DES), which was used from 1940 to 1960 to prevent miscarriages in pregnant women, has been shown to increase the risk for breast cancer. We now know that this risk is associated with epigenetic changes. One must wonder, then, whether these changes are passed down to the next generations, along with the increased risk.

Air pollution, especially from particulate matter in vehicular exhaust, also causes epigenetic changes that can lead to inflammation throughout the body. Benzene, which is found in gasoline and other oil-based fuels, leads to altered DNA methylation associated with leukemia. In our water supply, chlorination leads to by-products with names like trihalomethane, triethyltin, and chloroform, all of which can induce epigenetic changes in the genome. Many of these chemicals have been studied for detrimental effects on health. Rats with triethyltin in their drinking water suffered increased incidence of brain inflammation and swelling associated with increased methylation activities. Chloroform and the trihalomethane known as bromodichloromethane increased methylation in liver cells in a gene associated with liver disease.

Even benign substances we don't associate with such risks can have a hidden story in their production. Alarmingly, many Indian spices sourced from India have been found to be contaminated with heavy metals. The cause is likely the proximity of spice farms to smelting and mining operations and the resulting use of contaminated irrigation water. In 2013 alone, the FDA denied import of more than 850 spice shipments from around the world. To minimize such risks, U.S.-grown organic spices can be used safely, while care should be taken with those derived from India and China. Buying from reputable sources with known brand names can help. But one must be especially careful with spices obtained over the Internet or in unbranded, anonymous containers found, for example, in small neighborhood shops. In many cases, some specialty stores can obtain spices that bypass FDA inspection. While only about 2 percent of imported spices are found by the FDA to be contaminated, you significantly increase your odds of obtaining them when you buy unbranded spices from anonymous overseas sources.

Taken together, there is little doubt that a wide range of environmental toxins and pollutants can alter our epigenome, resulting in increased susceptibility to a host of different cancers (breast, liver, ovary, lung) and other diseases, including schizophrenia, diabetes, and heart disease. Each person's exposure is unique and different, which vastly complicates the problem. Some experts foresee the day when we will visit the doctor to get complete scans of our epigenetic alterations in order to determine our future risk for disease. Will we be increasingly using epigenetic-based drugs like HDAC inhibitors and RNA-based therapeutics to offset these risks and treat disease?

These scenarios are beginning to turn into reality. In this book we've offered an alternative you can pursue today, changing your lifestyle to mitigate risk, and perhaps in the future this approach, too, will be fine-tuned to specific epigenetic marks for disease. An even bigger question, based on studies like the ones cited here, is whether the epigenetic changes in adults living today will be inherited by the next generations tomorrow. Dr. Michael Skinner seems to have little doubt: "In essence, what your great-grandmother was exposed to could cause disease in you and your grandchildren."

Along these lines, it will be critically important to continue to be aware of how epigenetic modifications arise in response to environmental toxins and pollutants. This is the only way we can move forward, for the good of our own health and the health of generations as yet unborn.

ACKNOWLEDGMENTS

The new genetics has been one of the most rewarding subjects the two authors have ever written about, and because the ground being covered was vast, we have many people to thank. As long as the list is, every relationship was personal and personally gratifying.

Every book calls upon a publishing team, and *Super Genes* was fortunate to have such a superb one, beginning with our astute and encouraging editor, Gary Jansen. Also many thanks to others at Harmony Books who constituted and managed the working team: Diana Baroni, vice president and editorial director; Tammy Blake, vice president and director of publicity; Julie Cepler, director of marketing; Lauren Cook, senior publicist; Christina Foxley, senior marketing manager; Jessica Morphew, jacket designer; Debbie Glasserman, book designer; Patricia Shaw, senior production editor; Norman Watkins, production manager; Rachel Berkowitz and Lance Fitzgerald, foreign rights department.

We all know the pressures that book publishing is under today, and so a special thanks goes to the executives who must make tough decisions about which books to publish, including ours. Generous thanks to Maya Mavjee, president and publisher of the Crown Publishing Group, and Aaron Wehner, senior vice president and publisher of Harmony Books.

Our excitement over the breakthrough research in epigenetics was magnified by the Self-Directed Biological Transformation Initiative, a project that has been enormously fruitful thanks to a host of research collaborators. We offer our heartfelt thanks to all of you, including the following:

From the Chopra Center for Well-Being, Sheila Patel, Valencia Porter, Lizabeth Weiss, Wendi Cohen, and Sara Harvey and the entire staff.

The OMNI La Costa Resort and Spa, for generously accommodating our study.

Murali Doraiswamy, Arthur Moseley, Lisa St. John, and Will Thompson of Duke University.

Susanna Cortese at Massachusetts General Hospital and Harvard Medical School.

Eric Schadt, Sarah Schuyler, Seunghee Kim-Schulze, Qin Xiaochen, Jeremiah Faith, Milind Mahajan, Yumi Kasai, Jose Clemente,